API 1169 Pipeline Construction Inspector
Examination Guidebook

Craig Coutts and Paul Wilkinson

Wilkinson Coutts Ltd

© Craig Coutts and Paul Wilkinson 2019

This publication is in copyright. All rights reserved. Subject to statutory exception and to the provisions of relevant collective licensing agreements, no reproduction of any part may take place without the written permission of Wilkinson Coutts Ltd.

First published 2019

Typeset by Data Standards Ltd, Frome, Somerset, UK

Printed by Page Bros, Norwich, Norfolk, UK

A catalogue record for this publication is available from the British Library.

ISBN 978-1-9993459-0-7

The publisher has no responsibility for the persistence or accuracy of URLs for external or third-party Internet web sites referred to in this publication and does not guarantee that any content on such web sites is, or will remain, accurate or appropriate.

The publisher and the authors are not responsible for the reliability of any statement made in this publication. Data, discussions and any conclusions developed by the authors are for information only and not intended for use without independent substantiating investigation on the part of potential users.

Contents

Preface	vi
Chapter 1 How to use this book	**1**
Chapter 2 The API Individual Certification Programs (ICP)	**3**
Chapter 3 API ICP examinations: what to expect	**6**
3.1 Exam question format: what to expect	6
3.2 API exam question types	6
Chapter 4 The role of the pipeline inspector	**16**
Chapter 5 API 1169 Body of Knowledge (BoK)	**19**
5.1 What is different about the API 1169 BoK?	23
5.2 Why the excitement about inspector responsibilities, rather than technical details?	23
5.3 The content of API RP 1169	26
5.4 The CEPA/INGAA Practical Guide for Pipeline Construction Inspectors	34
Chapter 6 API RP 1169 and the CEPA/INGAA guide to inspection	**38**
6.1 API 1169 inspection requirements – new pipeline construction	38
6.2 The CEPA/INGAA guide for pipeline construction inspectors	51
6.3 Sample API 1169/CEPA/INGAA question sets	54
Chapter 7 Line pipe materials: API 5L	**77**
7.1 API 5L line pipe	77
7.2 Manufacture	79
7.3 Inspection issues	82
Chapter 8 The pipeline construction codes: ASME B31.4 and ASME B31.8	**84**
8.1 ASME construction codes – what are they?	84
8.2 ASME B31.8: what's the scope?	86
8.3 The liquid pipelines construction code: ASME B31.4	89
8.4 API 1169 exam questions	94
Chapter 9 The welding code: API 1104	**95**
9.1 The role of welding codes	95
9.2 So what about API 1104 itself?	96

9.3 API 1104 sample questions — 108

Chapter 10 Pipeline defects — 112
10.1 Indications and defects — 112
10.2 The acceptance criteria of API 1104 — 114
10.3 Pipeline surface distortions — 118

Chapter 11 Pipeline pressure testing: API RP 1110 — 123
11.1 Pressure testing – what's it all about? — 123
11.2 The pressure test itself: RP 1110 content — 126
11.3 Pressure test types — 130
11.4 RP 1110 sample questions — 134

Chapter 12 Federal pipeline regulations 49 CFR 192 and 49 CFR 195 — 138

Chapter 13 CGA best practice document 13.0 and INGAA crossing guidelines — 145
13.1 CGA best practice document 13.0 — 145
13.2 INGAA pipeline crossing guidelines — 149
13.3 Sample questions — 153

Chapter 14 Inspector health and safety responsibilities and the BoK — 161
14.1 API RP 1169's allocation of responsibilities — 161
14.2 H&S exam questions: treasure hunt — 164
14.3 H&S treasure hunt questions: question set 14.1 — 166

Chapter 15 OSHA health and safety regulations — 185
15.1 The problem with regulations: the unit of truth — 185
15.2 OSHA 29 CFR 1910 — 187
15.3 OSHA 29 CFR 1926 — 189
15.4 Trench excavation: 29 CFR 1926 subpart P — 191
15.5 Blasting and explosives: 29 CFR 1926 subpart U — 199
15.6 Operation of cranes and derricks: 29 CFR 1926 subpart CC — 202
15.7 Scaffolding and fall protection: 29 CFR 1926 subparts L and M — 208
15.8 Flammable liquids fire risks: 29 CFR 1926 subpart F — 209
15.9 Materials handling: 29 CFR 1926 subpart H — 212
15.10 Hazardous substances: OSHA 49 CFR 172 — 215
15.11 Welding and cutting safety: ANSI Z49.1 — 218

Chapter 16 The pipeline inspector's environmental responsibilities — 235
16.1 Inspector environmental responsibilities — 235
16.2 FERC document: *Wetland and Waterbody Construction and Mitigation Procedures* — 237
16.3 FERC document: *Upland Erosion Control, Vegetation and Maintenance Plan* — 241

16.4 The national pollution contingency plan: 40 CFR 300	244
16.5 Protection of navigable waterways	247
16.6 Protection of birds and endangered species	249

Chapter 17 The pipeline inspector's quality responsibilities — 259
17.1 Quality management: evolution or revolution? — 259
17.2 ISO 9000 and API Q1 — 261
17.3 The way forward – BoK tracker questions — 262

Chapter 18 Snippets — 270
18.1 Specialist pipeline terminology — 270
18.2 Pipeline mapping: the NPMS — 275
18.3 The longest pipeline in the world? — 276
18.4 Piping uphill, pipelines downhill — 278
18.5 Grizzly bears — 279
18.6 Pigs — 281
18.7 Mallard ducks — 283
18.8 Earthquakes versus pipelines — 284
18.9 The pipeline environmental impact assessment (EIA) — 287
18.10 How to learn from codes — 291

Chapter 19 Mock examination: 100 questions — 294

Appendix 1 Answers to sample questions — 319

Appendix 2 API SIRE exam guide: sample chapter — 329

Index — 349

Preface

Wilkinson Coutts Ltd are pleased to publish this industry guide to API 1169 Pipeline Construction Inspection. The book is intended for qualified overland pipeline inspectors and those preparing to sit for the API 1169 Certified Pipeline Inspector examination.

Overland pipeline inspection is a broad subject involving a wide variety of technical, safety and environmental disciplines. Large pipeline contracts with multi-national manufacture and arduous site conditions bring their own requirements of coordination, specification compliance and the quality of construction on site. Worldwide, the adoption of ASME, API and EN codes in many industries helps to bring a sound technical base to specifications and the activities of construction inspection.

The technical competence of individuals is one of the main criteria of effective pipeline construction. Pipeline inspectors need the ability to interpret and implement technical codes, specifications and statutory regulations across a range of disciplines. On-going improvement and certification of pipeline inspectors is a way to achieve this. This guide provides specific instruction for inspectors preparing for the API 1169 examination. This API programme is finding increasing acceptance in the pipeline industry as a way of demonstrating inspector competence.

Wilkinson Coutts are experienced providers of exam preparation training for many API certification examinations including API 510, 570, 653, 571 and 580. We also provide training for the API source inspector fixed equipment (SIFE) and rotating equipment (SIRE) certificates and the ASME plant inspector programmes in the UK and worldwide. Our course results speak for themselves and we achieve a 95% + first-time pass rate in most disciplines.

At Wilkinson Coutts we base our business on an interactive approach to training. We structure our courses to maximise delegate interaction and involvement rather than use passive slide presentations. We are also enthusiastic advocates of peer-to-peer (P2P) training techniques to maximise training effectiveness and time efficiency.

Contact us for our latest API, ASME and other technical course schedules for the integrity industry. We welcome feedback on our publications, so if you have any comments on the content of this book then please contact us.

Visit our UK website www.wilkinsoncoutts.com

For details of our associate company in Australia visit www.wilkinsoncoutts.com.au

95+% 1st Time API Pass Rate

Our specialised courses can be delivered in-house for companies wishing to train several staff, or on an individual basis on our public courses.

API Exam Preparation Courses:

- ✓ API 510 Pressure Vessel Inspector
- ✓ API 570 Process Pipework Inspector
- ✓ API 653 Storage Tank Inspector
- ✓ API 580 Risk Based Inspection
- ✓ API 571 Corrosion and Materials
- ✓ API 577 Welding and Metallurgy
- ✓ API Source Inspector Fixed Equipment
- ✓ API 1169 Pipeline Construction Inspector

Other Courses:

- ✓ ASME Certificate Plant Inspector Level 1/2
- ✓ API 579 Fitness For Service
- ✓ Pressure Testing Procedures
- ✓ Corrosion Awareness and Inspection
- ✓ AICIP Exam Preparation Training (Australia)
- ✓ DNV RP 103 Non Intrusive Inspection
- ✓ ASME PCC 2 Pressure Equipment Repair
- ✓ Pressure Relief Valves - Inspection & Testing

www.wilkinsoncoutts.com

www.wilkinsoncoutts.com.au

E: info@wilkinsoncoutts.com   T: +44 7753 808 738

Chapter 1

How to use this book

This book is intended to be a basic 'how to do it' guide to passing the American Petroleum Institute (API) 1169 Pipeline Inspector (PI) examination. API 1169 is the most recent API certificate examination to be introduced under the API Individual Certification Programs (ICP). It supplements the Source Inspector Fixed Equipment (SIFE) and Source Inspector Rotating Equipment (SIRE) exams that deal with the inspection of static (pressure) and rotating equipment during manufacture. This is known in various industries as either *works* inspection, *shop* inspection or, more recently, *source inspection*.

Pipeline construction inspection is a wide and multi-disciplinary subject, due partly to the location, i.e. on a construction site rather than at a manufacturer's works. This book limits its scope to those activities covered by the API 1169 examination, rather than attempting to be a full guide to pipeline design, manufacture and installation. As with all examinations, API certificate examinations cannot cover all the topics of a subject so you should treat this book as an exam guide rather than a full description of the subject of pipeline inspection.

You will not find this book to be a repeat of the detailed code information included in the examination Body of Knowledge (BoK). Candidates planning to take the exam have a long list of codes to refer to, so you will need these to read in conjunction with this book if you want to have a good chance of passing the exam. There are many code references spread throughout the book where you will be referred to a relevant code or recommended practice (RP) document relevant to the exam BoK. When you use these, make sure to use the current edition or amendment as given in the 'Code Effectivity List'. This gets updated from year to year so always check the version shown on the ICP pages of the API website www.api.org.

Will reading this book teach me to be a pipeline inspector?

Not entirely, but it is intended to give you a pretty good start. Pipeline inspection requires a combination of code familiarity plus sound (and wide) engineering knowledge, sharpened and tempered (at the same time) by a healthy dose of site construction experience. As with all exams, they don't make you an expert on all the aspects of a subject. You need to start off with the correct idea as to what the objectives of pipeline inspection actually are, and the best way to achieve them. This then makes your experience *count*.

When should I attempt the sample exam questions in the book?

This book contains lots of sets of sample questions, divided broadly into separate subjects. It's not really worth attempting these until you have read the chapter in this book that precedes each question set and studied the relevant code or document from the Code Effectivity List. Without this, the question sets will become little more than a guessing game. You won't learn much and probably won't get many correct either. Once you have read through the necessary material it is best to attempt the questions 'open-book', i.e. finding the answers in the relevant code or document. This will build up your familiarity with the documents; without this your chance of passing the exam are slim, at best. You can check your answers in Appendix 1 of this book. Each contains the code reference so you can check the source of each answer, whether you got it right or wrong.

Chapter 2

The API Individual Certification Programs (ICP)

API certificate examinations

American Petroleum Institute (API) certificate examinations come under the general banner of the API Individual Certification Programs (ICP). This is an expanding suite of examinations servicing the inspection/integrity part of the engineering industry.

Note these general features of these API (ICP) exams.

- API exams are **not for beginners**. They are not really aimed at trainees or new entrants to industry.
- API imposes **entry requirements** for candidates wishing to register for many of the examinations.
- **API examinations are difficult** – owing mainly to the long scope of the published body of knowledge (BoK) that forms the source material for each exam category. For some, this can amount to 2000+ pages of published codes and recommended practice (RP) materials.
- API examinations (as the name suggests) are written in US style, based on US codes, practices and examination style. This is consistent in itself, but can differ significantly from that used in other parts of the world. This is an important point that, depending on your background, may have a real effect on your ability to understand the programme material and pass the exam.

The API ICP scope

The following list shows the full scope of the API ICP examinations for 2018 onwards. At any time, the current list of ICP is shown on the API website (www.api.org/certification-programs). The content of this book fits in with the API 1169 Pipeline Inspector scope.

- API 510: Pressure Vessel Inspector

- API 570: Piping Inspector
- API 653: Aboveground Storage Tanks Inspector
- API TES: Tank Entry Supervisor
- **API 1169: Pipeline Inspector**
- API 571: Corrosion and Materials Professional
- API 577: Welding Inspection and Metallurgy Professional
- API 580: Risk Based Inspection Professional
- API 936: Refractory Personnel
- API SIFE: Source Inspector Fixed Equipment
- API SIRE: Source Inspector Rotating Equipment
- API QUTE: Qualification of Ultrasonic Testing Examiners (Detection)
- API QUPA: Qualification of Ultrasonic Testing Examiners (Phased Array)
- API QUSE : Qualification of Ultrasonic Testing Examiners (Sizing)
- API IA-Q1: Internal Auditor Q1
- API IA-Q2: Internal Auditor Q2
- API A-Q1: Auditor Q1
- API A-Q2: Auditor Q2
- API LA-Q1: Lead Auditor Q1
- API LA-Q2: Lead Auditor Q2

THE API 1169 programme scope

The BoK set by API for the API 1169 examination is large, diverse and widespread. It uses codes and published documents from seven different sources.

- American Petroleum Institute (API)
- American National Standards Institute (ANSI)
- Interstate National Gas Association of America (INGAA)
- Common Ground Alliance (CGA)
- International Standards Organisation (ISO)
- Federal Energy Regulatory Commission (FERC)
- US Code of Federal Regulations (CFR)

As an alternative, there are also documents from Canadian sources for delegates who wish to take the Canadian version of the examination. As you would expect, these documents have different structure, content and style. There is a fixed package of code documents for the API 1169 examination. This is shown in the Code Effectivity List in Figure 5.1

later in this book. These represent the BoK for the subject as decided by the API.

Registering for the API 1169 exam

This is a completely separate activity to participating in any training programme. It is each candidate's individual responsibility to pay and register for the API examination for the exam windows scheduled throughout the year. This is done via the websites of API (www.api.org) and its exam site contractor, Prometric (www.prometric.com).

API ICP exam training courses

There is no compulsory training required for candidates who wish to sit any of the API ICP exams. In theory, to become an API-certified inspector, all you have to do is apply to API, meet its acceptance criteria, book your exam (lasting between 3 and 7 hours depending on which ICP exam you are attempting) and then pass it. Some candidates can pass like this but, for most, unless you have full familiarity with the relevant codes (2000+ pages' worth for some ICP exams), you are unlikely to pass the exam and will need to prepare for the examination by enrolling on a training course. The training course will teach you about the subject matter covered, test you using mock exams and so on, and prepare you to take the API exam. Details of training courses in these API ICP subjects that we offer are available on our website (www.wilkinsoncoutts.com).

Important note: the role of the API

The API has not sponsored, participated or been involved in the compilation of this book in any way. The API does not issue past ICP examination papers or details of their question banks to any training provider, anywhere. API codes are published documents that anyone is allowed to interpret in any way they wish. Our interpretations in this book are built up from a 15-year record of running successful API training programmes in which we have achieved a first-time pass rate of over 90%. It is worth noting that most training providers either do not know what their delegates' pass rate is, or don't publish it if they do. The API sometimes publishes pass rate statistics – you can check the website (www.api.org) to see if they do – and what they are.

Chapter 3

API ICP examinations: what to expect

The API 1169 examination consists of 100 multiple choice questions comprising a mixture of 'closed-book' and 'open-book' questions. The required pass mark is typically 70%.

3.1 Exam question format: what to expect

API ICP exam questions are written in a specific way. If you haven't experienced them before they can look a bit strange, particularly if you are used to examinations written outside the USA. There's nothing actually *wrong* with the API question style, it simply has its own character that takes a little bit of understanding if you haven't seen it before. The main objectives of the exam questions are as follows.

- To test your knowledge of the code documents by asking you to recall statements included in them. This may also involve you understanding some of the basic principles underlying the verbatim wording of the codes.
- To test your skills in answering questions using only the wording used, rather than adding any imagined 'context' information of your own.

As with all the API ICP exams, the difficulty lies with the large number of documents related to API RP 1169 (*Recommended Practice for Basic Inspection Requirements – New Pipeline Construction*) that are covered in the body of knowledge (BoK). The codes and regulations are long documents containing lots of technical information and potentially thousands of exam question opportunities.

3.2 API exam question types

At first glance, API exam questions look like any others with different forms of wording and structure arranged in random order. On closer inspection, however, they fall into four distinct types, as illustrated in Figure 3.1. There are several variations on each of the types shown but, overall, you can identify all the questions as falling into one of these four types.

In order to increase the number of questions available to the question bank from a finite number of technical subject areas, the principle of *brother questions* is used. This consists of taking a question and changing the wording of the question and/or answer options slightly while staying within the same technical boundaries. The essence of the question remains the same, but it looks very different. It is not uncommon for a question on a common topic to spawn three or four different brother questions, which can appear in different parts of the exam paper. One piece of good news about brother questions is that if you know the correct answer to one question, you will also know the correct answer to its brothers, as long as you read the question carefully enough.

All the four types of exam questions can be made a bit more difficult by adding commonly used 'distractors' such as long-winded or redundant wording or negative statements that require you to think more carefully about what the question is actually asking for.

Type 1. Direct quote questions

Type 1 questions are basically about the word-for-word statements written in the code documents. The correct answer is the one that exactly replicates a string of words as they appear in the code. The other (incorrect) answer options sound like they could be correct but

- play with the meaning of words and phrases such as *equal to, less than or equal to, shall, should, likely, unlikely* and similar ones that closely define the answer
- contain a *not* statement, which changes the meaning of the answer
- contain a different word, or have a word missed out.

With type 1 questions, the way in which the word combinations are written in the code is considered more important than whether an answer is true for all possible engineering scenarios. It is assumed that this is an argument that has already been investigated when the code was written.

FIG 3.1
The four types of API exam questions

TYPE 1 QUESTIONS:
Verbatim wording quote

Q. The stresses imposed on a pipeline during a pressure test are effectively?

Ans. *Static*

TYPE 2 QUESTIONS:
Based on loose paraphrase or 'intent'

Q. The stresses imposed on a pipeline during a pressure test ignore?

Ans. *Fatigue conditions*

QUESTIONS ARE BASED ON SOURCE MATERIAL APPEARING SOMEWHERE IN THE CODES... LET'S SAY THIS IS IT:

The *objective* of pressure tests is sometimes misunderstood. It is part of the system of verifying the integrity of a pipeline but it has its limitations. The stresses imposed on a pipeline during a pressure test are effectively static; they impose principal stresses and their resultant principal strains. This means that what they test is the resistance of the pipeline only to the principal stress and strain fields, not its resistance to cyclic stresses (that cause fatigue), or the other mechanisms that have been shown to cause pipelines to fail. Hence the pressure test is *not* a full test of whether the pipeline will fail as a result of being exposed to its working environment and the incidence of steel pipeline actually failing catastrophically under a site pressure test is quite small . A pressure test **is not** a 'proving test' for pipelines that have not been properly checked for defects (particularly weld defects). It is also not a proving test for pipelines where unacceptable defects have been found – so that the pipeline can be somehow shown to demonstrate integrity, in spite of the defects.

TYPE 3 QUESTIONS:
Elimination or 'least wrong' answers

Q. A pressure test on a pipeline is?

(a) Unlikely to result in failure
(b) A test for fitness-for-purpose
(c) A test for all defects
(d) A proving test

Ans. (a) It may not always be correct in all contexts, but it is the *least wrong* based on the code text it was sourced from.

TYPE 4 QUESTIONS:
Based on general knowledge that may not be directly traceable to the form of words in the code

Q. Hydrostatic tests are?

Ans. *Preferred to pneumatic tests for safety reasons*

Here is an example of a type 1 question from the API 1169 scope.

Q. The pipeline inspector's relationship with contractors, suppliers and vendors

Pipeline inspectors (PIs) are expected to establish a professional business relationship with contractors, suppliers and vendors. Inspectors must

(a) Help contractors select the manner to perform contracted work ☐
(b) Not direct nor supervise the contractor's work ☐
(c) Check the contractor's work at all times ☐
(d) Evaluate the contractor's work at all times ☐

The answer is (b). Depending on the activity, you could argue that (a), (c) or (d) could be correct in some contexts. If, for example, the contractor is performing a specific repair to a pipeline spool to meet a non-conformance raised by the PI, then it may be a good idea for the PI to witness the repair and evaluate the results. This, however, is not the point. The question text does not state this scenario, so it is incorrect (in the API exam world) to assume it.

The answer comes from no more than a straight quote from section 4.4 of RP 1169, which states that *Inspectors must respect their position and not direct nor supervise the contractor's work*. Notice that the first five words of this sentence do not appear in the text of the exam question. Instead, they are replaced with a verbatim sentence from the beginning of the paragraph, i.e. *Inspectors are expected to establish a professional business relationship with the contractors, suppliers and vendors*. This sounds so good that its effect carries over to the answer options, encouraging you (perhaps) to continue this spirit of closeness with the contractor and choose one of the wrong options (a), (c) or (d).

Type 2. Paraphrase questions

In type 2 questions the 'correct' answer has been chosen from the general text as it appears in the code, but has some words changed to produce a *paraphrase*. This may be deliberate or accidental – it doesn't matter which. Owing to the indirectness of the word-match, type 2 questions are more likely to be suitable for the closed-book part of the API 1169 exam.

In order to catch out people who like to rush into answers, it is not unusual for words like *rarely*, *mostly*, *generally* or *likely* to be added in somewhere to make things a little more difficult to interpret.

Here is an example of a type 2 question and a 'brother question' developed directly from it.

Q. Final coating inspection

Site coating of pipeline sections generally takes place

(a) In the manufacturer's works ☐
(b) After lowering-in to the pipe trench ☐
(c) After pressure testing ☐
(d) Before pressure testing ☐

The answer is (c) (RP 1169 section 7.13.4). This paragraph does not specifically say that final site coating is performed after pressure testing. Nor does RP 1169 (section 7.18) *Hydrostatic pressure testing requirements* specifically say that testing is performed after lowering-in to the trench. This paragraph does, however, come after these covering lowering-in (section 7.14) and backfilling (section 7.15) so the inference is that it happens *after* these activities. Hence options (b) and (d) are not the correct ones.

Option (a) refers to the coating applied in the manufacturer's works, while the question is about coating on site, hence eliminating (a). This leaves the correct answer as (c). Now here is a type 2 brother question developed directly from the one above.

Brother question: final coating inspection

During site construction of a buried pipeline, which of these activities is carried out immediately before lowering the pipe into the trench?

(a) External coating ☐
(b) Active cathodic protection ☐
(c) Welding ☐
(d) Pressure testing ☐

The answer is (a). This is developed from the previous question, although looking nothing like it in presentation. It revolves around exactly the same point, i.e. that site coating of the circumferential welds is performed after welding but before lowering-in to the trench. Pressure testing is done after that. Option (b) active cathodic protection is a nonsense answer – it involves an impressed current system that is set up after backfilling. Note the word *immediately* that is included in the question. The purpose of this is to eliminate (c) welding from being the correct answer. Welding is obviously done before coating of the welds is done, but normally not *immediately* before.

Type 3. Elimination questions

These are strange things that offer a good way to check candidates' knowledge of a subject and their ability to analyse the wording of a question. An answer is deemed to be correct because it is *less wrong* than the other options. One feature of these questions is that they are very general and lack much qualifying information. Too much qualifying information would negate the correctness of the designated answer to apply to different engineering situations, so it is missed out, leaving it totally general. These questions do not appeal to many engineers, because *least wrong* does not necessarily mean *right*.

Here is an example of a type 3 question.

The mechanical property of a steel known as ductility is

(a) The ability of a material to be formed ☐
(b) Linked to toughness ☐
(c) Related to material strength ☐
(d) Elasticity ☐

The answer is (b). This is a typical technical question chosen because it can be used in many of the API ICP examinations. It's a bit too detailed for the PI role but that doesn't matter. The origin of the answer comes from the fact that the ductility of a metal is what provides blunting at crack tips when a crack is trying to propagate, hence increasing the resistance of the metal to crack propagation (i.e. its *toughness*). No doubt metallurgists would argue that some of the others may be true in some situations; however, this is an exam question for PIs, not metallurgists.

Type 4. 'General knowledge' questions

Thankfully, these are fairly rare, as most API 1169 examination questions are sourced from the content of the code documents in the BoK. Occasionally, however, some facts are considered sufficiently widely known to be considered general knowledge that an inspector should possess, by virtue of past training and experience. Type 4 questions are included to trip up candidates with very poor experience, but should not prove difficult at all to anyone with a reasonable level of experience and background knowledge in engineering subjects. Here is an example.

During site assembly of an above-ground fabricated pipeline on site, most defects are found in

(a) High-stress areas of the parent material ☐
(b) Small bore nozzles where welding access is difficult ☐
(c) Welds ☐
(d) Areas of dissimilar materials ☐

The answer is (c). This is because the question-setter has decided that *welds* are where most defects will be found, so thinks you should believe that also. You might argue that small-bore nozzles can contain welds, or that it is perfectly possible to have a small-bore nozzle welded to a high-stress area of the parent material. You are thinking too deeply – this is a valid API general knowledge question and the answer is *welds*. It may even be true.

Now, the distractors

All the four question types can be changed (and brother questions created) by the application of a few *distractors*. Here they are.

Distractor A: the '*not*' question
A 'negative' appears in the question statements or the answer options. These are difficult for candidates for whom English is not their first language.

Distractor B: the '*almost correct*' answers
These are plausible-sounding incorrect answers that either represent common mistakes or misapprehensions, or just simply sound correct, but are not. They catch out guessers, skim-readers and lazy people.

Distractor C: questions lacking context
These are questions that lack background context information that would fully describe what the question is actually asking, thereby enabling a fully reasoned answer. The objective is to tempt you to make up your own context in line with your own experience, or lack of it. One of the principles of ICP exam questions is that candidates should answer the question as it stands, however incomplete it looks, and not construct their own imaginary context to it. Have a look at the following example.

Q. Inspector responsibilities

In most cases a PI's main responsibility is to

(a) The stakeholders in the pipeline project ☐
(b) A pipeline company ☐
(c) Local and national enforcement authorities ☐
(d) All of the above ☐

This is a good example of a question that is seriously lacking in context information. It gives no clue as to how the project is structured or who is working for whom. The only clue in the question is the qualification saying '*in most cases*'. This specific form of words identifies it as either a verbatim quote or paraphrase question. The first stop in answering this question correctly is to avoid the famous *all of the above* option (d). If you look at this option for long enough, your brain, clever thing that it is, will read into options (a), (b) and (c) what it wants to see, justifying (d) as being the best answer. It does this because it doesn't like uncertainty – it would *like to believe* that (d) is the answer. Putting this another way, unless it is absolutely and justifiably certain that (a), (b) or (c) are *incorrect*, it will default to seeing what it wants in (a), (b) and (c), therefore concluding that (d) is the right answer. It's an illusion of course – but that's what it does.

Coming back to what is the correct answer, the words in the question appear in RP 1169 (section 4.2) entitled *Owner/operator representative*. Following on, it says that the PI works (in most cases) for a pipeline company, so that's the answer – (b). Yes, this is a crude system that ignores any context where the PI could be working for the pipeline contractor, pipeline manufacturer, regulatory authority or any other party. These other parties can, and do, employ inspectors. In the question-setter's eyes of course, none of this matters – the answer to the question fits the wording of the code text and that's that.

The lesson from this?

The lesson from this holds good for all API ICP examinations as they all use the same philosophy. The lesson is

- As an exam candidate, do not read any context into an exam question if it isn't written there.

If you didn't get this at first reading, here are a couple of examples.

If a question were to read 'Inspectors are expected to report *deficiencies* to...' *don't assume* the question refers to material deficiencies, welding deficiencies or any other specific *technical* deficiencies. It doesn't say that, so they could be environmental spills, incorrect personnel certificates, unqualified machinery drivers, missing site weather reports or a myriad of other non-technical things that would change the choice of the correct answer. Instead, look for the answer option that best describes how you would report *any deficiency at all*. The answer will either appear in the text of a code somewhere, if it is an open-book question, or be logical when you look at it from a closed-book perspective, having not added any of your own imagined context along the way. Don't be surprised if the only feasible answer option fitting this criterion is one that you consider weak – that just shows you it's a poor-quality question and you have to expect some of those.

The next example is particularly relevant to the API 1169 exam owing to the way that 50% of its BoK is about non-technical subjects such as quality, safety and environmental issues. Here it is along with the answer options.

Q. Environmental and pollution control requirements

Regarding environmental and pollution control, a pipeline inspector should

(a) Ensure the owner/operator has procedures for compliance in this area ☐
(b) Report to the relevant authorities/jurisdiction if procedures are inadequate ☐
(c) Understand where their job function involves *covered* activities ☐
(d) Pay particular attention to the protection of *uncovered* excavation equipment during storms or high winds ☐

Let's try something different. Instead of there being only one 'correct' answer (according to the question-setter), assume for a moment that they are all correct. Your task is to rate them in order of correctness starting with the most correct and ending with the least correct, as you see it. While you are doing it, think about a real pipeline construction site, with its remote, rough terrain and inhospitable climate. Try this now before continuing to the next paragraph.

. . . . .Try it now. . . . .

What did you get for your answer? It's a fair bet you rated the options something like (b), (a), (d), (c) or perhaps (a), (b), (d), (c). Sadly, the most (only) correct answer is (c), hidden away in RP 1169 (second paragraph of section 6.1). If you didn't rate (c) first, and many people wouldn't, you were successfully misled by the poor paraphrase wording, and a bit of subtle misdirection about the meaning of the word *covered*. You have to watch out for strange/awkward/unfair/tricky/misleading/bad questions like this.

Chapter 4

The role of the pipeline inspector (PI)

The PI's job

For site PIs, their job takes place at the pipeline construction site. This is different to the situation for an API source inspector (SI) whose role is in the manufacturer's works to ensure the problems never make it to the stages of site assembly and operation. The idea behind both of these roles is that deficiencies are identified before they get big enough to cause a problem. In practice, this can never be 100% achieved; however, you can contrast this to what would happen if an inspector was *not* present.

Difficulties

Any inspection is about monitoring the work and performance of others so it is rarely a comfortable procedure. As an inspector it is normal for you to be dealing with experts who know more about the subject than you do. Against this awkward background, inspection involves finding problems, criticising or influencing work carried out. For this reason, not everyone will see your input as being either useful or constructive. You should not expect therefore that you will always (in fact *regularly*) be working within your own comfort zone. It just doesn't work like that.

Management activities

As a PI, part of the job entails managing various site contractors and their sub-contractors and sub-manufacturers 'down the chain'. They do the work and have to be encouraged to get it right.

Another part is the need to manage the expectation of *your client*, which will generally be the pipeline operating company. Clients generally know *broadly* what they want in terms of project procedure and completion, but their high-level view sometimes means that they cannot assimilate all of the technical detail. Here are the results.

As a PI, or any other type of SI for that matter, you can expect your client to occasionally

- **disagree** on which part(s) of a technical or project specification are the most important, and their view might change day-to-day
- **misinterpret** codes, standards and regulations to fit their own views of what they would like them to say
- **delay** awkward decisions, perhaps in the hope they will go away
- **accept** things that are not ideal if the alternative would be to delay the project schedule.

You can see all of these, most days, in a large pipeline project.

Technical activities

Unlike some inspector roles, the job of the PI has both technical and project responsibilities. Duties consist of

- **involvement** in the execution of the project inspection and test plan (ITP), mainly by performing quality surveillance
- **witnessing** tests (of many types) and reporting the results
- **accepting** or rejecting test results and reporting compliance with specifications.

These are mainly technical activities relating to compliance of the construction project against its specified requirements. This technical part of the PI's role includes witnessing tests. There are many types, but common ones are

- material tests, for mechanical properties of pipeline parent material or welds
- non-destructive tests/examinations (NDE) of pipeline welds and coatings
- pressure tests
- functional tests, on valves for example.

Tests normally follow a written procedure to make sure they are done properly. Some tests are just simple visual and dimensional checks against weld maps, weld specifications or construction/general arrangement (GA) drawings. Others are more complicated, requiring specialist test equipment in the laboratory or workshop.

Working to acceptance criteria

Most technical tests witnessed by a PI have some specification or code-specified acceptance criteria that have to be met. Material properties, NDE and pressure tests are all good examples of these. Judging results against acceptance criteria is therefore one of the key technical roles of the PI. Remember, however, that that is not the *whole* role – tests must be seen in the context of the pipeline construction project as a whole, which is sometimes the difficult part.

Chapter 5

API 1169 Body of Knowledge (BoK)

Whichever way you look at it, the API 1169 BoK is a bit of a monster. Subject matter is drawn from

- 10 standards/documents from the American Petroleum Institute (API)/Canadian Standards Association (CSA), American National Standards Institute (ANSI), Canadian Energy Pipeline Association/Interstate National Gas Association of America (CEPA/INGAA) and International Standards Organisation (ISO), forming the closed-book section of the examination questions

plus

- 12 more from the US Code of Federal Regulations (CFR), Federal Energy Regulatory Commission (FERC) and United States Codes (USC) (or their Canadian equivalents), forming the open-book part of the exam

then, as if that wasn't enough

- A couple of American Society of Mechanical Engineers (ASME) technical standards to be used 'for background information'.

Figure 5.1 shows the full Code Effectivity List – this is updated regularly and available with the official BoK on the API website (www.api.org). The examination consists of a total of 100 questions drawn from the closed-book and open-book sections. Unlike most of the other API ICP examinations, both sets of questions are contained in the single 100-question exam paper, with only the open-book code list available for reference.

FIG 5.1
API 1169 Exam Code Effectivity List

Listed below are the effective editions of the publications required for the API 1169 exam up to February 2019. Please consult the API website for the up-to-date list for February 2019 onwards. Please be advised that API documents are copyrighted materials. Reproducing copyrighted documents without permission is illegal.

CLOSED-BOOK REFERENCES

- **API Recommend Practice 1169**, Basic Inspection Requirements – New Pipeline Construction, 1st edition, July 2013, reaffirmed August 2018
- **API Recommended Practice 1110**, Pressure Testing of Steel Pipelines for the Transportation of Gas, Petroleum Gas, Hazardous Liquids, Highly Volatile Liquids, or Carbon Dioxide, 6th edition, February 2013, reaffirmed August 2018
- **API Q1**, Specification for Quality Management System Requirements for Manufacturing Organizations for the Petroleum and Natural Gas Industry, 9th edition, June 2013 (with Errata 1, February 2014, and Errata 2, March 2014). Sections 3–5 only
- **ANSI Z49.1**, Safety in Welding, Cutting, and Allied Processes, March 2012, Chapters 4, 5, 6 and 8
- **CEPA Foundation /INGAA Foundation**, A Practical Guide for Pipeline Construction Inspectors, March 2016
- **CGA**, Best Practices, Current edition
- **INGAA**, Construction Safety Guidelines Natural Gas Pipeline Crossing Guidelines, Version 1, June 2013 CS-S-9 Pressure Testing (Hydrostatic/Pneumatic) Safety Guidelines, September 2012
- **ISO 9000:2015** Quality Management Systems – Fundamentals and Vocabulary, 3rd edition (confirmed in 2009). Definitions only
- EITHER **API Standard 1104**, Welding of Pipelines and Related Facilities, 21st edition, September 2013 Addendum 1 and Addendum 2 (May 2016) (Sections 3–11 only, excluding appendices) OR **CSA Z662-15**, Oil and Gas Pipeline Systems, June 2015 (chapters 1, 2, 4, 6, 7, 8, 9 and 10)

OPEN-BOOK REFERENCES

Applicants may use either the US Federal Regulations or the Canadian Federal Regulations

Please check the API website for pdfs of the below references

US References	Canadian Equivalents
Code of Federal Regulations (CFR) **49 CFR 192**, *Transportation of Natural and Other Gas by Pipeline: Minimum Federal Safety Standards:* Article 7, Subpart E, Subpart G, Article 614 & Article 707 **49 CFR 195**, *Transportation of Hazardous Liquids by Pipeline:* Article 2, Article 3, Subpart D, Article 310 & Article 410 Safety	
Safety	
29 CFR 1910, *Occupational Safety and Health Standards (OSHA):* Article 119, Subpart I (excluding Article 140 & Subpart I Appendices), Articles 145–147 (excluding Appendices) & Article 184 **29 CFR 1926**, *Safety and Health Regulations for Construction (OSHA):* Subpart C, Article 62 (excluding Appendices), Article 102, Article 152, Articles 250–251, Article 451, Articles 500–501, Article 601, Subpart P, Article 902 and 914 & Article 1417 **49 CFR 172**, *Hazardous Materials Table, Special Provisions Hazardous Materials Communication, Emergency Response Information, Training Requirements, and Security Plans:* Article 101 (excluding appendices)	**Canada Occupational Health and Safety Regulations (COHS)** Parts III, IV, X, XI, XII, XIV, XV and XIX **Transport Canada** Transportation of Dangerous Goods Regulations : Parts 1.4, 2 (excluding Appendices 1 & 3–5), 4 & 6 Environmental

Environmental	
33 CFR 321, *Permits for Dams and Dikes in Navigable Waters of the United States* **40 CFR 300**, *National Oil and Hazardous Substances Pollution Contingency Plan:* Subparts A & E **Federal Energy Regulatory Commission:** Office of Energy Projects *Wetland and Waterbody Construction and Mitigation Procedures*, May 2013 *Upland Erosion Control, Revegetation, and Maintenance Plan*, May 2013	**Canadian Environmental Protection Act, 1999 (S.C. 1999, c.33)** Sections 3, 64–65 & 90–99 **Fisheries and Oceans**, *Land Development Guidelines for the Protection of Aquatic Habitat:* Section 3 **Canada Water Act** (R.S.C., 1985, c.C-11): Part II **Canadian Energy Pipeline Association (CEPA)**, *Pipeline Associated Watercourse Crossings*, 4th edition, November 2012
Migratory Bird Permits (50 CFR Part 21) Subpart B	**Migratory Bird Convention Act, 1994 (S.C. 1994, c.22) (F)** Sections 4–6 & 12
33 USC Chapter 9: Protection of Navigable Waters and of Harbor and River Improvements Generally Subchapter I Articles 401, 403, 403a, 404, 407	**Navigation Protection Act (R.S.C., 1985, c. N-22)** Sections 2–14 & 21–26
Endangered Species Act of 1973: Sections 3, 4, 7, 9, 10, 12	**Species at Risk Act (S.C. 2002, c. 29)** Sections 2, 32–39 & 56–64

Note: All examination questions are based on the materials listed above. The American Society of Mechanical Engineers (ASME) documents below are recommended for general knowledge but not required for the exam. All exam-related information contained within ASME documents can also be found in API RP 1169 and CEPA/INGAA's Practical Guide for Pipeline Construction Inspectors.

American Society of Mechanical Engineers (ASME)

> **B31.4**, Pipeline Transportation Systems for Liquids and Slurries, 2012 edition Chapters I–III and V and VI only
> **B31.8**, Gas Transmission and Distribution Piping Systems, 2014 edition, General Provisions and Definitions and Chapters I–IV & VI only

5.1 What is different about the API 1169 BoK?

Quite a few things – API 1169 is unique in its approach and content compared with the other API ICP exams. The three main differences are as follows.

- It is a qualification for construction inspectors working on *installation sites*, rather than in manufacturing works.
- It is based predominately (perhaps 90%+) on **inspector responsibilities, duties and proceedings** rather than the technical details of pipeline design and construction.
- There is no official API-published exam study guide, as there is for the other source inspection examination programmes (SIFE for fixed equipment and SIRE for rotating equipment).

Don't expect any great hidden agenda behind these points. Their origin is simply in the practical differences between a construction project involving onshore pipelines extending over long distances in remote areas and traditional pressure equipment, vessels, pipework and components manufactured within a single self-contained plant site.

Although pipeline sections (the popular term is 'line pipe') are rolled, welded, coated and tested in a fabrication shop, the final assembly, welding and coating is done on site along the pipeline route, as the sections are connected up to form a continuous pipeline. Construction may last years, crossing many different areas and terrains, each bringing its own difficulties and challenges.

5.2 Why the excitement about inspector responsibilities, rather than technical details?

The main reason for this is simply that of *necessity*. Long-distance pipeline construction in remote places is a world apart from the closely controlled ordered environment of the fabrication works. Multiple contractors work at a distance from their head office co-ordinator and technical support. Extended logistics and timescales can result in a shifting sands of contractor personnel, with planning and training often performed 'on the job' to respond to problems and changes as they arise. The net result of all this is an environment in which the scope of the PI's role vastly exceeds that of the traditional source inspector (SI) making periodic or occasional visits to a manufacturing works.

In a site situation, the PI becomes an integral part of the construction project, rather than an occasional impartial observer. The scope widens from a purely technical role in assuring contract compliance to include project planning and procedural matters, with parallel responsibility for various safety environmental and communication matters.

It is this wide scope of the PI's responsibility that is the subject of API RP 1169. There it is in the title: *Recommended Practice for Basic Inspection Requirements – New Pipeline Construction*. With this breadth of scope, an inspector's *detailed* technical knowledge takes second place to the need to understand the inspection requirements of the wider project scope, while knowing *where to find* technical detail when it is necessary.

Fortunately, this imposed limitation on the PI's technical knowledge level is helped along by the nature of pipelines themselves. Pipeline designers no doubt think pipelines are items of great technical complexity, but to others they are just about the most straightforward bits of engineering around. There are a few different types of construction, various ways to weld one piece of line pipe to the next, no doubt, but whatever method is used is repeated, with little variation, hundreds or thousands of times as the pipeline is assembled along its length. Yes, there are other bits and pieces involved (pig catchers, flowmeters, isolation valves and the like but they are not *that* complicated). In addition, operational damage mechanisms such as high temperatures, chemical processes and fatigue are less arduous than experienced on refinery or process plant, so material choice and design are less complicated. This is all good news for PIs.

So what?

If you have followed the argument so far, the contents list of RP 1169 should now make sense to you. Following some years of integrity problems with onshore pipeline projects, RP 1169 was written as a specific response to industry problems, centring on the importance of the inspection activity. More recently (2016) a support document, the CEPA/INGAA foundation document *A Practical Guide for Pipeline Construction Inspectors*, was adopted as a support document in the API 1169 examination BoK. More about this later.

Back to the BoK

Now that you know the rationale behind the BoK, have a look at Figure 5.2. It gives a pictorial view of the breakdown of the subjects involved. Remember that all the subjects shown are related to the inspector role and activities that lie within each one, rather than the essential technical

API 1169 Body of Knowledge (BoK) 25

FIG 5.2
API 1169 Body of Knowledge
– Subject/Exam Breakdown –

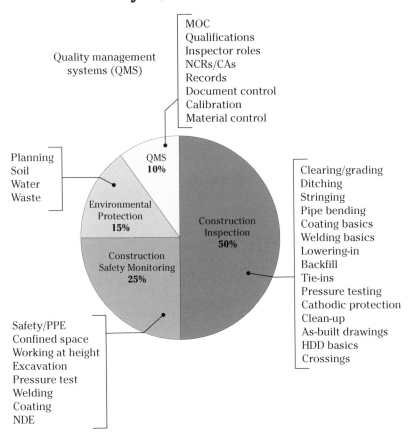

REMEMBER THAT THESE ARE MAINLY
SITE INSPECTION ACTIVITIES
(not at the manufacturing works)

content of the subject itself. As you can see, only 50% of the scope involves physical inspection of the pipeline site construction itself. 25% is about the monitoring of safety-related activity and 15% linked to controlling the environmental effects of excavating a large trench and putting a pipeline in it, then reinstating landscape afterwards. The final 10% relates to operation of the quality management systems – mainly that bit of it installed to keep the inspection activities under control, rather than the myriad of other things that quality assurance (QA) can no doubt be applied to, if you have the time and enthusiasm.

Note the subject breakdown that makes up each segment of the BoK sections in Figure 5.2. These limit the technical boundaries of the API 1169 BoK scope – particularly important in safety and environmental scopes, which, left unrestricted, could go on forever.

5.3 The content of API RP 1169

Look a bit deeper at the BoK and the contents page of RP 1169 and you can see, perhaps unsurprisingly, that the BoK is not that different from the breakdown of the content of the RP 1169 document itself. Figure 5.3 shows the content breakdown of RP 1169. Read this clockwise starting at the 12 o'clock position and it shows the chapter numbers and rough page allocation for the included subjects. Construction inspection, safety and environmental-related inspection are all there in RP 1169 chapters 4, 5, 6 and 7, just in a bit of a different order from that used in the formal BoK listing.

RP 1169: have a closer look

For this, you need to look at the document itself, starting with the title and contents page followed by a quick scan through the content. The key points are as follows.

- RP 1169 is **not a technical document**, like many of the API codes and their RPs which are devoted to engineering detail. Instead, it is a document *allocating responsibilities and activities*, i.e. prescribing who does, or doesn't do what. That's its role.
- Assume it **is all (and only) about inspection activities** and you won't go far wrong. You can add *ensuring, monitoring, confirming and compliance-checking* to the list if you like; they probably differ slightly in the degree of 'doing' involved, but all mean basically the same.
- There is no long list of technical terms and definitions in RP 1169

FIG 5.3
RP 1169 content breakdown

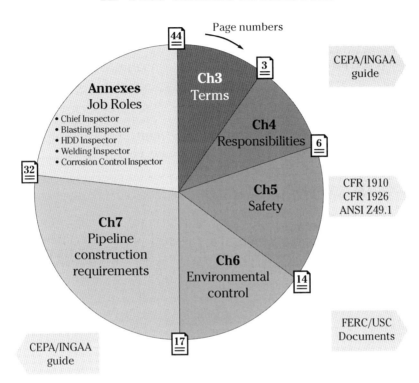

The CEPA/INGAA guide adds detail to the RP 1169 inspector roles and responsibilities during site construction

Like this:

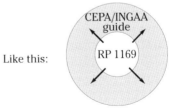

chapter 3, which is where these things normally appear in profusion in API documents, just a couple of simple definitions.
- Lists of **inspection responsibilities are everywhere**. Although RP 1169 chapter 4 is dedicated to the subject, the allocation continues equally as enthusiastically in all the other chapters also. Remember, it *is* a book of responsibilities.

Moving on: the PI role

We now know that throughout the 50 or so pages of RP 1169, the role and responsibilities of the PI are prescribed just about everywhere. This is, of course, the problem. The extent of the prescription is so widespread that it would be easily possible to construct 2000 or more examination-style questions based on the pure verbatim wording of all the clauses and sub-clauses. Change the wording of each one around a bit and you have 2000 more, and so on. Based on this, your chance of getting a pass mark from 100 (even well-written) examination questions (all closed-book remember) becomes next to zero at best. With statistics and luck in your favour, they become slim, to 25%.

Fortunately, you have an ally in this seemingly unequal competition: the idea of *context*. If you can understand the role of the PI as it fits into the context of the real-world pipeline construction programme then the task becomes so much easier – even manageable. All you have to do is to understand what PIs *are* and why they are there on construction sites, and multiple questions on PI roles and responsibilities will fall straight out, almost answering themselves.

To save time, try it this way: start by considering what the PI is **not**.

Pipeline inspector? – no, not me

Figure 5.4 shows what the PI is *not*. Of the six bullet points listed, these roles are all taken by other parties, either as a consequence of their contractual position in the pipeline project or their role as a technical specialist or sub-contractor of some sort, which is much the same thing. All these other parties will have more knowledge than the PI about the 'doing' aspects of the bullet points. Hence they do the doing and the instructing, leaving the PI to play second fiddle. That's the reason why, among many others of equivalent meaning, clause 4.4 of RP 1169 dictates that *Inspectors must not direct nor supervise the contractor's work*.

Keeping this in mind can give you a major insight into the nature of many of the API 1169 exam questions. Used correctly, it's an effective

API 1169 Body of Knowledge (BoK) 29

FIG 5.4
What a Pipeline Inspector certainly is NOT

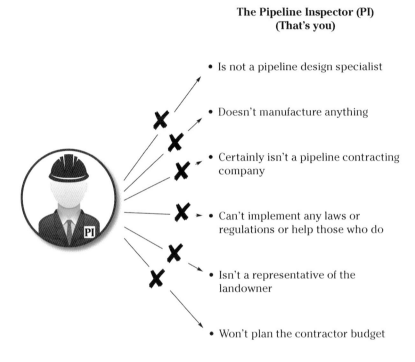

The Pipeline Inspector (PI)
(That's you)

- Is not a pipeline design specialist
- Doesn't manufacture anything
- Certainly isn't a pipeline contracting company
- Can't implement any laws or regulations or help those who do
- Isn't a representative of the landowner
- Won't plan the contractor budget

SO: The PI can't be expected to have the specialist knowledge these parties do.

SO: What does a PI do all day on site? **See Fig 5.5**

way of eliminating more than half of the incorrect answers from consideration, clearing the way for consideration of the more viable answer options. All we have to do now is to address the question about what the PI actually *does* do all day. For that we will soon need to look at Figure 5.5.

Hello, I'm the pipeline inspector

The job of the PI has both good and bad points. On the good side, the nature of most pipeline construction projects is such that a PI enjoys the advantages of contractual back-up for their activities. That means that the construction contract legitimises their right in being there (a good start) and their role in monitoring, participating and interfering in all manner of site activities. This is frequently much more than a SI can expect when doing ad-hoc or regular inspections in a manufacturer's works, where the contractual link can be tenuous or poorly defined.

Another positive aspect of PI work is the technical discretion that arises from the lack of definition of the PI role in the ASME B31.4/31.8 codes specified for pipeline construction. There is no direct equivalent of, for example, the ASME Authorised Inspector as in the ASME VIII pressure vessel code, where the standards the Authorised Inspector has to work to are set out in technical detail, leaving not too much room for discretion. Coupled with the technical simplicity of pipelines and their technical codes, all of this makes the technical part of the PI's role just a little easier. Much the same argument applies to the exam content. In a recent (2018) amendment to the API 1169 BoK, the pipeline technical construction codes ASME B31.4 and B31.8 were relegated to 'background information only' status. This eliminates their purely technical content from forming either open- or closed-book exam questions.

Figure 5.5 sets out what the role of the PI actually is. As a representative of the pipeline owner/operator, the PI *ensures*, *monitors*, *confirms*, *checks*, *verifies* and *reports* that everything (just about) on the project is going as it is supposed to. The remit covers procedural, safety and environmental and general QA issues, as well as the compliance of the construction with its purely technical requirements. Let's look in a bit more detail at these items.

Technical compliance
The PI monitors compliance with the technical requirements of the pipeline as it is assembled, tested and completed. Compliance is checked against contract specifications, technical codes and 'line drawings' showing the pipeline engineering features, fixtures and fittings. In this way, the role is similar to that of the traditional shop-based SI, except that it is performed at the construction site.

Quality (QA) monitoring
RP 1169 (section 4.3) charges the site PI with a larger-than-expected responsibility for assuring the overall quality of the construction

FIG 5.5
What the Pipeline Inspector does

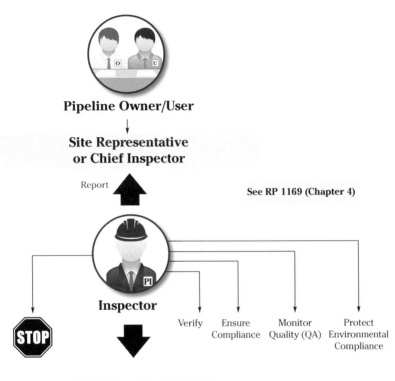

project. This increases the contribution from the more passive role of the shop-based inspector monitoring compliance with an inspection and test plan (ITP) arranged by others, to an active role with day-to-day involvement on how the plan works and the quality of the output. Inevitably this requires active involvement in regular project and QA meetings, acting as part of the project team rather than just an impassive observer.

Project planning participation
The PI is not a project planner – that is the job of the project planner. Notwithstanding this, the extended chronological nature of a pipeline project means that the inspection, monitoring and verifying activities that the PI has to do cannot be separated from the incidence of when and where these activities will happen. This links the PI firmly into the planning and scheduling activities happening on site – once again, a much more involved role than a shop-based SI. Now you see why the PI needs to be a permanent fixture on site rather than just an occasional visitor.

Safety representative also?
Partially, yes. RP 1169 devotes a healthy eight pages to the role of the PI in monitoring, verifying and assuring (all those words again) that safety-related procedures and activities are performed properly on the construction site. As before, the *responsibility for compliance* rests with the main site contractor but there is strong emphasis on the PI's active participation in the activities. A key responsibility arising from this is

- It is the PI's job to *report* unacceptable safety practices (see RP 1169 (5.1))

To be able to do this, the PI needs a wide awareness of hazards and risks that can arise from all the activities that occur on site. This brings the two US legislation documents into the equation. Known as 29 CFR 1910 and 29 CFR 1926, they provide extensive coverage of safety activities relating to lifting, excavating, materials handling, hazardous chemicals, radiological risks and similar. A related document (ANSI Z49.1) covers risks in the welding and cutting of metals. This presents the PI with a wide scope of monitoring; easier to list than to carry out in practice. For API 1169 exam purposes, it means that the questions are of fairly broad character. The individual areas listed in RP 1169 section 4.3 give you a good guideline.

Finally, the PI as a cost controller?
Officially, even practically, the PI will never be the cost planner or cost controller for a pipeline construction project. It is a separate discipline, initiated and managed several stages up the chain of contract hierarchy by people traditionally suspicious and resentful of informal interference from below. This doesn't of course prevent RP 1169 (section 4.4) from sneaking bits of it into the role of the site PI. The PI is allocated a role in monitoring and exerting some control over the cost issue, while accepting that the PI is naturally not the major player. The rationale for this resides in the fact that technical or procedural decisions made by the PI can all too easily affect costs. It is more common for these to raise costs rather than to reduce them (surprise, surprise). This potential effect of PI involvement varies with the type of cost that you consider.

- **Bulk material costs**. These are the least able to be influenced by PI decisions and reports. Apart from the occasional damaged and rejected pipe length, bulk material costs will remain much the same.
- **Schedule delays**. These are the largest because any inspection activity that results in work hold-up or repetition has the potential to cause time delays. On site projects, time delays mean cost increases owing to the highly labour-intensive nature and long sequential timescale of the activity. Increased equipment hire costs also occur, at an alarming rate. For this reason, the PI needs to take a proportionate view of the knock-on effects of inspector requests and activities. This is one direction RP 1169 is coming from when it mentions the role of the PI as *participating* in cost control.
- **Systematic technical deficiencies**. RP 1169 is careful to differentiate individual technical deficiencies from situations where deficiencies, whatever they may be, are recurring and so indicate there may be a systemic problem with some activities. In such cases these are escalated to the status of non-conformance (NC). Because of the time involved in reporting, discussing and solving these, there will nearly always be some cost implication. They will either be solved immediately or kicked down the road to reappear later in the project, which they do. Problems with automated welding and non-destructive evaluation (NDE) procedures can act like this, resulting in the need for retrospective testing, even re-work, of long sections of pipeline.
- **Hydrostatic pressure testing costs**. Pressure testing is an expensive business, requiring special contractors, water supply/disposal consents and safety procedures to be implemented over a large physical

area. Mistakes by the PI (approving incorrect tests, accepting rejectable defects etc.) can result in escalating repeat testing costs, plus the costs of excavation and repair if defects appear later in use.

This list of duties outlines, as seen in RP 1169, the rules and responsibilities of the construction site PI. RP 1169 lists them only as principles, however – to add detail we need to look at its partner document in the BoK, the CEPA/INGAA document *Practical Guide for Pipeline Construction Inspectors*. This was only added to the BoK in 2017 with its role being to support RP 1169, adding real-world procedural details for the PI to follow.

5.4 The CEPA/INGAA Practical Guide for Pipeline Construction Inspectors

Let's call it the CEPA/INGAA guide for short. This document was recently published by CEPA and INGAA, the two main North American pipeline associations, in response to a continuing problem with the quality of gas and liquid pipeline construction. Their stated objective is to achieve a step change in construction quality via the role of the PI. It is rapidly gaining acceptance in North America and being adopted by most stakeholders in the pipeline industry.

At 130 pages long this guide acts as a substantial back-up to the content of RP 1169. It starts by confirming the surveillance role of the PI as initially set out in RP 1169 (section 4.1) and expands to incorporate an *expectation of performance* for the PI's role on the construction site. In doing this it expands the PI's role a bit beyond the margins of RP 1169 with some additional duties and responsibilities. The main content of the guide is to break the activities of the PI role into the 12 sequential steps of site construction programme. Figure 5.6 shows the breakdown.

The CEPA/INGAA guide is a book of lists

Most of the content of the CEPA/INGAA guide takes the form of *lists* of inspection-related duties and activities. Some of these are quite long and not that interesting to read but are a good representation of the various documents and activities you will find on a real pipeline construction site. Within the context of all these lists, the activities of the PI are set out in the following series of steps, which are to

FIG 5.6
The CEPA/INGAA
Guide for Pipeline Construction Inspectors

This provides detail on the duties of the Pipeline Inspector during the phases of a pipeline construction project.

– IT'S A BOOK OF CHECKLISTS –

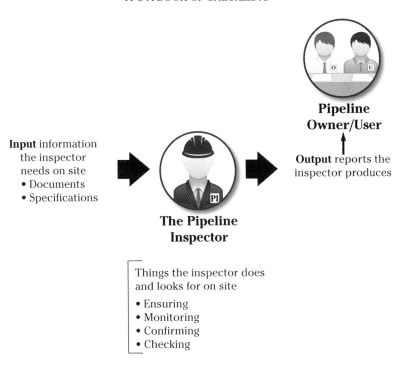

Pipeline Owner/User

Input information the inspector needs on site
- Documents
- Specifications

Output reports the inspector produces

The Pipeline Inspector

Things the inspector does and looks for on site
- Ensuring
- Monitoring
- Confirming
- Checking

The 12 phases of the construction project

Survey › Clearing & Grading › Stringing › Field Bending › Ditching & Excavation › Welding › Coating › Lowering in › Backfilling › C.P. › Pressure testing › Clean-up & restoration

- observe
- monitor
- assess
- evaluate
- verify
- decide
- resolve

At first glance, you can be excused from concluding that all these activity words mean very much the same. Look a bit closer though and you can see that they are actually all slightly different, progressing in chronological order (more or less) from the beginning of the list to the end.

Links to the API 1169 BoK

You can get some good insights by looking at how the content of the CEPA/INGAA guide relates to the published BoK for the API 1169 examination. The list of subjects given in the BoK (which states that there will be at least one exam question from each of the listed subjects) broadly mirrors the content of both RP 1169 itself *and* the CEPA/INGAA guide. The three documents are closely linked together in their content, even though there are not many actual cross-references included in them. For exam purposes it therefore makes sense to treat these as a 'matched set', as that essentially is what they are, albeit in different disguises.

One inherent difficulty you will encounter with the CEPA/INGAA guide is that lists are not easy things to learn from. Sequential reading of lists will, after the first few minutes, not help you retain much of the information stored within them. This is made worse by the repetition of information across many of the lists, providing almost a subconscious instruction for your mind to switch off its learning capability and wait for something more interesting to come along. You can get a partial solution to this by picking at selected items from the lists (documents required, tests to be witnessed etc.) that coincide with a mention of the activity in the equivalent section of RP 1169. It's a painstaking job but is just about the only way to pick out the list-based items that really matter.

To conclude, it is a fair bet to assume that more than 50% of the 100 closed-book API 1169 exam questions will have their origin in either RP 1169 itself or the CEPA/INGAA guide. Even questions that are sourced from the wording of one of the other codes on the Code

Effectivity List will have their rationale hidden away in RP 1169 or the CEPA/INGAA guide somewhere. Think of these two documents as the equivalent to the API study guide published for some of the other ICP examination scopes and you won't go that far wrong.

We will look in more detail at the content of the CEPA/INGAA guide and its multiple lists in Chapter 6.

Chapter 6

API RP 1169 and the CEPA/INGAA guide to inspection

6.1 API 1169 inspection requirements – new pipeline construction

As discussed in Chapter 5, RP 1169 is a document allocating responsibilities for inspection-related activities on the pipeline construction site. Looking back to Figure 5.3 you can see the breakdown of RP 1169 into its component chapters and their parallel to the API 1169 examination body of knowledge (BoK). If you look at the actual content of the BoK on the API website you can see how it is mainly expressed in terms of the chronological activities that make up the pipeline construction project, i.e. right of way (RoW) surveying, trenching, lowering-in, backfilling and testing. At first glance, the content of RP 1169 itself doesn't mirror this exactly, being divided more into functional responsibilities such as those for safety, environmental and pipeline construction. Read slightly between the lines, however, and the influence in the BoK becomes apparent – all the content *is there*, just arranged in a slightly different way.

Some content of RP 1169 is worth looking at separately. This will help your understanding of how it fits in with the other codes in the BoK and therefore the part it plays as a source of examination questions. Chapter 4 of RP 1169 is all about the principles of the role of the pipeline inspector (PI). We saw this broken down in Figures 5.4 and 5.5. It is also careful to define the limits and boundaries of the job: look at chapter 4 of RP 1169 and you can see the limitations of the PI when dealing with

- media/PR issues
- contact with landowners

- planning and scheduling.

Conversely, it *reinforces* the role of the PI in the areas of QA, safety and environmental compliance, taking it outside the scope of the works-based source inspector (SI). This is a key difference, establishing the PI's responsibilities in strictly non-technical areas of the project. Chapters 5, 6 and 7 of RP 1169 are devoted to itemising these responsibilities in more detail, providing cross-references to other BoK codes to fill in the detail.

What are the PI's safety responsibilities?

Perhaps more than you would expect – and they are listed in chapter 5 of RP 1169. While there is nothing particularly new in this chapter, it provides clear confirmation of the wide extent of safety responsibilities heaped into the PI role.

RP 1169 chapter 5: personnel and pipeline safety requirements

It is clearly stated in the API 1169 exam BoK that 25% of the 100 exam questions are about safety. Chapter 5 of RP 1169 therefore goes into significant depth, dividing the requirements into no less than 20 sub-sections. At the centre of the activity (section 5.2) lies the participation of the PI in the Job Safety Assessment (JSA). The JSA is the site-specific responsibility of the owner/operator and acts as the top-level guidance document for safety management of all the parties and activities, present and planned, on the construction site. Figure 6.1 shows the situation.

Sections 5.3–5.19 of RP 1169 give you a summary of the safety-related activities and confirm the role of the PI in each. Note that there are few (if any) significant differences of the perceived role of the PI in any of these – they are simply phrased as general statements of where the PI's knowledge should lie. Don't expect much detail here about *what* exactly it is that the PI should know about what, it just relies on blanket phrases like

- *Inspectors should be knowledgeable of* ...

or

- *Inspectors should have a basic knowledge of* ...

and leaves it as that. Note how this is all couched in RP-style non-mandatory *should* terms, rather than mandatory *shall* requirements used

FIG 6.1
The RP 1169 Job Safety Assessment (JSA)
– Look how wide it is –

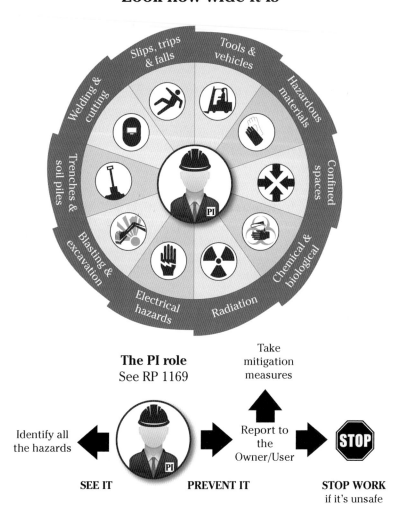

in publications with official code status. Overall, this can result in API 1169 exam questions being less well defined and more woolly than comparable API 510/570/653/SIFE exams.

To support the limited text of RP 1169, chapter 5 cross-references specific classes of the two main safety documents in the BoK

- 49 CFR 1910: *OSHA Occupational Safety and Health Standards*
- 49 CFR 1926: *OSHA Safety and Health Regulations for Construction*

There's not that much technical difference between these two documents, just in their method of subdivision and, presumably, legal interpretation in the USA. Their style of writing is what you would expect to find in a set of legal regulations: comprehensive, heavily subdivided into sub-clause and sub-part numbers, and of course rather uninspiring, given the nature of their content.

Do they define the PI's role?

No they don't, because they were written to apply to everyone, without specific reference to individual roles in any contract hierarchy. Responsibilities are mainly referenced to those of *the employer*, with occasional reference to *the contractor*, an umbrella term for just about everyone else working or present on the site. The PI is not identified as a separate role so, strictly, these Code of Federal Regulations (CFR) regulations do not add to, or qualify, the description of the PI role in RP 1169. The nearest you will get is confirmation of the responsibilities of the site employer (pipeline owner/operator) who employs the PI as a significant (but not the only) part of the team responsible for implementing things on site. On balance, this ties the PI firmly into the structure of responsibility for safety matters, hence the 25% of safety-related questions in the API exam.

The PI as environmentalist?

According to the API 1169 BoK, 15% of the exam questions are about environmental protection related to water, soil, vegetation or wildlife in some way. To support this, the BoK contains several US documents from the CFR, FERC (Federal Energy Regulatory Commission) and USC (United States Codes). Equivalent versions cover the Canadian version of the BoK. The good news is that all of these exam questions are open-book, with the codes available to view as pdf documents during the exam. The downside is that this volume of literature is large and spread out over all the documents, so answers are difficult to find by random searching. The exam questions are not particularly *difficult*, just difficult to answer correctly by educated guessing.

FIG 6.2
Pipeline Inspector environmental responsibilities

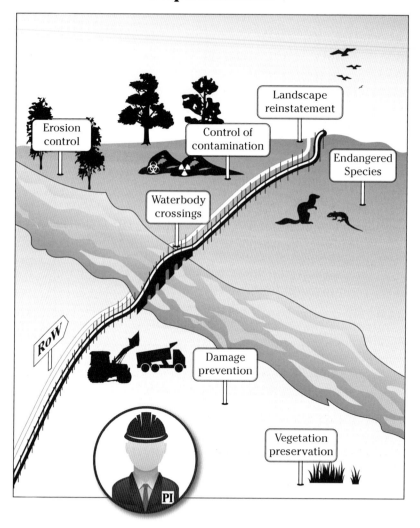

This environmental responsibility can cause difficulty for PIs whose main background is based mainly on mechanical engineering aspects of engineering construction. As we saw in Figure 5.3, the content of RP 1169 gives a clear remit to the PI to participate in this area. Figure 6.2 summarises the situation, with the PI responsibilities drawn from the itemised content of chapter 7 of RP 1169.

PI: prevention or reporting?
On a real pipeline construction site, a PI needs to be optimistic to believe they can act to *prevent* environmental problems, at least on a major scale. Site procedures for environmental and pollution control (that is the formal term used) are set by, and under the control of, the pipeline contractor and proposed, discussed and agreed well in advance of the PI's involvement. They will have involved environmental specialists, experts and non-experts of all shades of opinion and viewpoint on soil, vegetation, waterbodies (rivers and ponds), flora, fauna, fish, animals and all manner of endangered creepy crawlies that panic at the first sight of a pipeline.

Surrounded by all of this, it is the PI's job (see RP 1169 section 6.1) to *ensure compliance* with all the various procedures and any legislative requirements that lie behind them. Given the wide spread of activities going on, this is a difficult task as the PI clearly cannot be in all places at once. Practically then, the bulk of the PI's role tends to revert to *reporting* of non-compliances with procedures. That's just how it works.

For API 1169 exam question purposes, two particular topics can be singled out as slightly more important than the others

- the reporting of environmental contamination (section 6.7) and
- issues relating to water crossings (section 6.5).

Some typical questions on these are included in the samples at the end of this chapter.

As with the other chapters of RP 1169, chapter 6 cites cross-references to other codes and regulations. Not all of these are included in the exam BoK, however, so you need to be selective about looking these up. Overall, it is unlikely that more than one or two exam questions will be sourced from the cross-references cited in RP 1169 chapter 6.

In summary, the PI is not *in charge* of environmental compliance on a pipeline construction site. However, it is the PI's job to know *where to fit in* to the monitoring and reporting of compliance. That presupposes a basic level of knowledge of what the environmental compliance issues are, and the codes and regulations that cover them.

RP 1169 chapter 7: pipeline construction inspection

At 26 pages, this is the longest chapter of RP 1169. Looking back at Figure 5.2 you can see that construction inspection activities account for a straight 50% of the examination questions. Set out in the 26 pages is a long list of those activities of the pipeline construction project of which the PI *should be knowledgeable* (there's those words again) in order to meet the expectation of their pipeline owner/operator client. Looking back also at Figure 5.6 you can see how this list of sequential project activities is mirrored in the content of the CEPA/INGAA pipeline inspection guide. This covers the same activity steps but sets out in more technical detail the real-world actions, documents, test reports and so on raised by RP 1169 as lying within the PI's scope of responsibility. We can look at each of these in a bit more detail.

First, the verification of qualifications
The verification of key construction personnel qualifications and certification falls squarely within the role of the PI. The requirements are quite widespread, with qualification/certification requirements scattered across the ASME, ANSI and CFR documents, as well as RP 1169 itself. In addition, these codes cite qualifications pertaining to the American Welding Society (AWS), the American Society for Nondestructive Testing (ASNT) and NACE International, for technicians involved in welding, NDE and coatings. It is not unusual on a typical pipeline construction project for there to be 20–25 different sets of qualifications and certifications to be checked across the key personnel involved. Clearly, to be able to handle this, the PI needs to know

- **which activities** *need* recognised qualifications and certification as mandated by the various construction codes, and which do not
- **who** will actually be doing these activities, so they can be approached for verification (most are unlikely to come forward unannounced)
- **what form** the qualification and certifications take, and the rules regarding attainment levels, recognition of equivalents and renewals.

As if that were not enough, annexes A to E of RP 1169 set out preferred qualification and/or experience requirements for five additional and project-specific roles, namely the

- Chief Inspector
- Blasting Inspector
- Horizontal Directional Drilling (HDD) Inspector

FIG 6.3
The additional inspector roles of RP 1169 (annexes A to E)

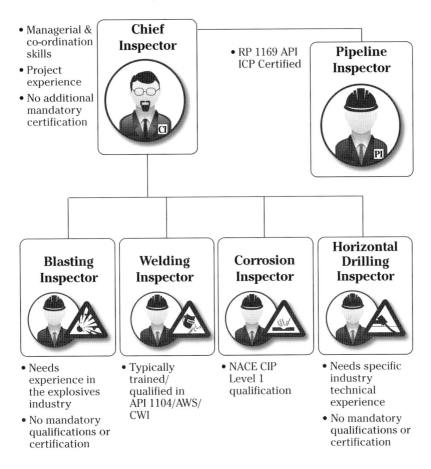

- Welding Inspector
- Corrosion Control Inspector.

Figure 6.3 shows these roles.

Note that these annexes are classified as *normative* additions to the document, meaning that they are relevant to the application of the document but not exactly part of it (if you see what I mean). Either way, they define project positions that the PI clearly does not occupy, so you can expect an exam question or two along those lines.

PI involvement with right of way (RoW) requirements

This is altogether a bit of a strange one. Subjects such as pipeline route selection, land surveying and the legal agreements covering these things are likely to be at the periphery of the field of competence of most PIs – nevertheless you can see some responsibilities set out in RP 1169 sections 7.3.1 to 7.3.6. The *Locating and marking requirements* in section 7.4 are a bit more practical and feasible for the PI to understand the route marking and communication system requirements set out in the referenced code 49 CFR 195 and other pipeline industry documents. A few PI responsibilities carry over into RP 1169 section 7.5, *RoW preparation requirements*, mainly to do with monitoring the clearing of vegetation, grubbing (removal of tree stumps) and grading (flattening) of the RoW to ensure it is done properly.

Ditching and excavation

The PI's responsibilities here are split between the safety aspects of excavating and routing the trench and the physical requirements for it to receive and support the steel pipeline in the correct way. Looking at the BoK, it is clear that this is one of the more important part of the PI's role, owing to its ongoing nature throughout the timescale of the pipeline project.

For API 1169 exam purposes, it is worth looking up the cited cross-references in OSHA 1926, 49 CFR 195 and 49 CFR 192 and the Common Ground Alliance (CGA) best practice document – all included in the exam BoK. Exam questions typically centre on safety features of trenches such as sloping, benching and shoring requirements and the attendant excavation risks using heavy machinery and rock blasting.

Pipe handling and stringing

This is project terminology for collecting pipeline lengths from the manufacturing works, bringing them to the site and laying them out (stringing; see Photo 6.1) end-to-end along the pipeline RoW. RP 1169 section 7.7 defines the PI's duties at this stage as mainly checking the physical condition of the pipeline section for defects, coating and marking. This is a straightforward inspection exercise, with acceptance criteria given in ASME B31.4 (434.4).

Pipeline components such as valves and flanges are generally inspected at this stage. There are a lot of ASME and API technical codes and RPs around covering acceptance criteria for these components, but inspection activities for these components are mainly covered in the SIFE (source inspection of fixed equipment) API examination so don't expect more than a few in the API 1169 PI exam.

Photo 6.1   Stringing (photo courtesy 123RF)

Inspection for out-of-code distortion after cold field-bending (RP 1169 section 7.9) is an important point and a good subject for exam questions. Although pipes are not usually bent through tight radii during cold bending on site, distortion can occur, producing wrinkles and misalignment of weld joints, which can be a cause for rejection.

Pipeline and waterbody crossings
A new pipeline RoW crossing near an existing pipeline or a waterbody such as a river, lake or wetland is considered, in the pipeline world, a common source of potential problems. Mechanical (or blasting) excavation of trenches can cause mechanical damage to existing installations and there are strict rules for such activities crossing, or even just being located near to, natural waterbodies.

Sections 7.11 and 7.12 of RP 1169 define the PI's involvement in these activities. There is not much actual engineering inspection involved here; instead, the main PI activities are described as

- checking crossing permits
- safety procedures where the RoW meets roads and railroads
- environmental procedures involving waterbodies
- documentation and reports.

In practice, verifying procedures, documentation and reports will be the main part of the PI's role in this part of a pipeline project. The INGAA and CFR documents in the API 1169 BoK contain any technical detail that the PI needs to know. Note that in the API 1169 exam, questions can be chosen from either open- or closed-book BoK areas. This can sometimes make these questions a little difficult if you are not that familiar with them.

Cathodic protection (CP) corrosion control

PIs cannot expect to be experts in CP systems. The principle is simple enough but the practice is a complicated mix of science and experience, and is thus the reserve of a few specialist companies. To do the job properly, CP technicians need specific qualifications and experience; you can see these set out in RP 1169 annex E.

The PI as coating inspector?

Just about all cross-country pipelines that are externally coated for corrosion protection have their coating applied in the manufacturer's works. Inspecting this activity is therefore the role of the works SI rather than the PI. The problem is, however, that the pre-coated pipe sections have to be welded together on site so a length of 300–400 mm at each end of each pipe section is left uncoated in the works. This then has to be coated on site, once post-weld heat treatment (PWHT) and non-destructive evaluation (NDE) have been completed. The coating is done by specialist coating contractors and completed before the assembled pipe is lowered into its trench.

The well-established spark (or 'holiday') test is used to check the coating for discontinuities such as nicks, breaks and pinholes. It is common for the PI to witness (but not perform) this test. Checking of pre-coating surface preparation is also important. The spark test does not test for adhesion and if the surface is not prepared to the correct degree of cleanliness and surface profile the coating will soon peel off. This causes a real problem as, by then, the pipeline is installed in the trench and backfilled, and so cannot easily be re-coated.

Let's all watch the lowering-in

Yes, this is in the PI's remit according to RP 1169 (section 7.14). The assembled pipeline is lowered into the prepared trench using multiple cranes (or excavators fitted with lifting fittings) to limit the amount of bend of the pipeline as it goes in. Even with this, it bends a surprising amount. In practice, most of the PI's inspection activity takes place

Photo 6.2 Pipe bending (photo courtesy 123RF)

before the lowering-in operation commences. The main things to check are

- safety of the lifting operation (the CFR codes), including the machines and all the lifting fittings, slings, hooks etc.
- preventing damage or distortion to the pipeline by incorrect slinging or collisions
- the suitability of the trench bottom to accept the pipeline without undue damage or distortion.

All of these demand a fairly constant schedule of visual inspections prior to and during the lowering-in procedure. On long overland pipeline routes this can go on continually for weeks or months, making it very difficult for the PI to always be in the right place at the right time. Realistically, it is rarely possible to witness everything and the PI will need to place some reliance on the pipeline contractor doing the work properly. Examples of pipe bending and lowering-in are shown in Photos 6.2 and 6.3.

Backfill and clear-up
Correct backfilling of the trench involves using the correct backfill material, to the correct depth and compaction of the material once it is

Photo 6.3 Lowering-in (photo courtesy 123RF)

backfilled. Some pipelines require softer padding material to cover the pipeline before a coarser material is used to finish off. Again, these involve straightforward visual inspections by the PI.

Line cleaning and hydrostatic testing
RP 1169 sections 7.16 to 7.18 cover these activities. They are sequential, with internal cleaning of the pipeline of soil and debris being a necessary step before filling and hydrotesting with clean water. Pipeline pigs are used to flush cleaning water through, prior to this filling. Several run sequences may be needed depending on the pipeline arrangement and the overall length to be pressure-tested.

Hydrostatic testing has sufficient safety-related and technical issues to be covered by a separate document in the exam BoK: the grandly named API RP 1110 *Pressure Testing of Steel Pipelines for the Transportation of Gas, Petroleum Gas, Hazardous Liquids, Highly Volatile Liquids or Carbon Dioxide*. There is also supporting safety-related information in 49 CFR 192 (J) and 49 CFR 195 (E), plus the INGAA document CS-S-9 (only 10 pages) which is in the closed-book question part of the examination. The technical aspects of pipeline hydrostatic testing are covered separately later in this book (Chapter 11).

Finally, the inevitable documentation

Section 7.20 of RP 1169 specifies clearly the minimum list of documents that the PI is expected to complete. This list combines technical documents with those about more procedural matters such as work/progress reports, work permits and the like. Further mandatory reporting requirements are cited from 49 CFR 195 and 49 CFR 192, although these documents are aimed predominantly at the site owner/operator.

Other RP 1169 inspectors – who are they?

You could be forgiven for thinking that the wide role of PI we have been discussing is the only inspection job of relevance in RP 1169. Fortunately, there are others. Annexes A to E of RP 1169 give five designated roles, each with the appropriate knowledge and skills requirements. Figure 6.3 shows them, with the main RP 1169 duties for each. Note that it's the principle of RP 1169 not to specify mandatory requirements, instead leaving them as '*should*' or '*recommended*' requirements. Note also that not all these five inspection positions need formal qualifications and certification; relevant experience and familiarity with the activities of the role are sometimes enough.

6.2 The CEPA/INGAA guide for pipeline construction inspectors

A Practical Guide for Pipeline Construction Inspectors, published in 2016, is the joint product of two North American pipeline organisations – CEPA (Canadian Energy Pipeline Association) and INGAA (Interstate Natural Gas Association of America). Both organisations have gained increased relevance in the industry due, in part, to the rapidly increasing number of gas pipelines for shale gas developments. To date there are more than 300,000 miles of gas pipeline in the USA and 200,000 miles in Canada, with many new routes in planning and construction.

The 2016 CEPA/INGAA guide was assembled with the clear objective of improving the quality of inspection practices throughout the lifetime of a pipeline construction project. It is therefore written with inspection and the duties of the inspector firmly in mind. It is based on the activities and experience of real pipeline projects and so represents a good benchmark of pipeline industry 'best practice'. Figure 6.4 shows its structure.

FIG 6.4
The CEPA/INGAA guide

	Its structural/engineering content		
Section	Activity	Structural/Engineering content/refs*	CEPA/INGAA guide table nos
7	Survey	–	Tables 15–28
8	Clearing & grading	–	Tables 29–46
9	Stringing	ASME B31.4/31.8 marking and damage assessment	Tables 47–59
10	Field bending	ASME B31.4/31.8 misalignment and bending limits	Tables 60–67
11	Excavation & ditching	–	Tables 68–82
12	Welding	API 1104 Welding: techniques and defect acceptance criteria	Tables 83–90
13	Coating	–	Tables 91–98
14	Lowering In	ASME B31.4/31.8 construction requirements	Tables 99–109
15	Backfilling	ASME B31.4/31.8 pipeline damage assessment	Tables 110–123
16	C.P.	–	Tables 124–132
17	Pressure testing	API RP 1110 Hydrostatic pressure testing. CS-S-9 pressure test safety	Tables 133–152
18	Clean up & restoration		Tables 153–167

*Limited to API 1169 exam BoK refs only

API RP 1169 and the CEPA/INGAA guide to inspection 53

The CEPA/INGAA guide was added to the API 1169 BoK starting from the 2017 examination. It supplements the RP 1169 document by adding the *detail* of PI activities to the roles and responsibilities of the PI raised by RP 1169 itself. The level of detail is high and it has clearly been written from operational knowledge of real overland pipeline projects.

In a few areas, the role of the PI as seen by the CEPA/INGAA guide is extended farther than it is in RP 1169. There is no conscious contradiction as such, but it does add some additional responsibilities around the margin. Most of these probably come from the reality that pipeline activities take place on remote sites where roles sometimes have to be stretched among the people available at the time. Examples of this are the PI's role in some of the personnel-related matters such as project planning and reporting of personal violations on site, which are covered in chapter 6 of the guide.

The overall role of the PI is summarised in the guide using the family of words *monitoring*, *observing*, *assessing*, *evaluating*, *verifying*, *deciding*, *resolving*, *reporting* and *documenting*, to ensure that everything is done properly.

Document structure

Following a few pages of introductory notes forming chapters 1–5 of the guide, chapter 6 sets out the following eleven sub-sections, outlining the responsibilities of the PI.

- The authority vested in the PI
- Code of conduct to follow
- PI role in site safety
- The role in monitoring conformance (and non-conformance)
- Environmental responsibilities
- The task of monitoring construction quality
- Administration role in the contract
- Contribution to managing records (this is a major item)
- Verifying personnel qualifications and certification
- Equipment calibration
- Duty to report site incidents

These are accompanied by the first in a long series of long itemised lists (the guide numbers them as tables) each containing lots of bullet points referring to items of information documents or actions.

The problem with the lists

The existence of the long itemised lists (there are 141 of them, no less) provides the first indication that while the CEPA/INGAA document provides an excellent guide to industry, it is not at all exam-question-friendly. Any itemised list containing more than about six points becomes almost impossible to learn from. This is because, for your brain to absorb information, it has to continually be able to link packets of information with which it is presented to bits that come before and after. In this way it can build up a picture of a story, filling in the gaps as it goes to create a consistent file of information to add to your memory bank. Long itemised lists don't fulfil this criterion, so your mind treats them as it would a long list of random numbers or playing cards. It'll remember the first few in the hope that something more related and interesting might be coming along, then, when it doesn't, lose interest and promptly forget the lot. In fact it has no need to forget because there was nothing remembered that needs forgetting. Try it and see – read one of the tables, wait 2 minutes, then try and recall what was there, and you will find that you can't.

What's the solution to the list problem?

There isn't one, because the way in which you absorb information is hard-wired into you and you can't change it. The good news, however, is that you actually *don't need* a solution. This is because long list formats are equally as difficult for exam question-setters to source sensible closed-book questions from as they are for you to remember. Short of choosing some random point or form of words from a couple of the 141 lists (containing an average of say 12–15 points each), there is little definitive enough to ask questions about. Overall then, it is likely that any exam questions drawn specifically from the CEPA/INGAA guide will be sourced from the general statements of inspection objectives and principles in the short chapters 1–5 or, at a push, chapter 6, about the responsibilities of the PI given in the text commentary rather than the bulleted lists that follow. Questions of this nature could probably also be sourced from RP 1169 itself.

6.3 Sample API 1169/CEPA/INGAA question sets

Your first sets of sample exam questions are taken solely from the content of RP 1169 and the CEPA/INGAA guide discussed in this chapter. They are split into five sets of questions as follows.

API RP 1169 and the CEPA/INGAA guide to inspection

- Question set 6.1: RP 1169 – responsibilities
- Question set 6.2: CEPA/INGAA guide – responsibilities
- Question set 6.3: CEPA/INGAA guide – general
- Question set 6.4 CEPA/INGAA guide – inputs and outputs
- Question set 6.5 CEPA/INGAA guide – monitoring activities

As introductory questions, they stick to the specified role of the PI and the knowledge base needed to do the job. Some are concerned with the limits or boundaries of the PI's role, i.e. what the PI's responsibilities are *not* (as we looked at in Figure 5.4). Taken together, the subjects broadly represent the 10%/50%/25%/15% split of subjects in the published BoK; have a look back at Figure 5.2 for a reminder of this.

Expect to see specialist terms, abbreviations and acronyms in some of the sample questions. The pipeline industry uses a lot of these, so if you don't know what they are, look them up in the terminology section of the codes – they are all there somewhere.

The way to learn

Questions drawn from RP 1169 and the CEPA/INGAA guide are in the closed-book question part of the exam, meaning candidates do not have these documents available as reference material. The best way to pick up the information required is to prepare by going through RP 1169 first, highlighting those areas that define

- what the PI *should know*
- subjects on which the PI *should be knowledgeable*
- topics the PI *should have a basic understanding of*.

These are little more than different ways of saying the same thing, so don't read any great difference into their meaning. There are a lot of these in RP 1169 – not surprisingly, as that is what the document is all about.

In these early sample question sets, the objective of most of the questions relates mainly to matters of principle (about what the PI does and doesn't do) rather than clever interpretation of verbatim wording given in the codes. This will be more prevalent when we move to the open-book style questions.

Finding the correct answer

There is little to be gained by guessing the answer to a question, getting it right or wrong, and then moving on to the next one. Have a guess at the question answers closed-book by all means, but then once you have checked the answer, look up the code clause cited in the answers and see *why* the correct answer is what it is. Sometimes it comes from a simple statement in the text and other times the answer may be inferred from the general meaning of a sentence or paragraph. It can also be, given the long-standing style of API ICP questions, the *most correct* (or least wrong) answer option.

When you are looking at the code clause words that define a correct answer, look also at the related text above and below it. This will hold the answer to a similar related question residing in the API question bank. It is a badly kept secret that each section of the API exam question bank contains multiple questions drawn from the same code areas and related closely to each other. You can think of these as brother (or sister) questions, differing only by an answer-changing term such as shall, should, not, maximum, minimum etc. Related questions like this are essential to enable several slightly different versions of each examination to be used in the multiple exam locations worldwide during the 2–3 week exam time-window.

Your test performance

You should aim to get at least 70% for your first attempt at each of the tests. You will only be able to do this consistently if you have read carefully through the preliminary chapter of this book *and* have related what it is saying to the content of the CEPA/INGAA guide. That means looking at the documents themselves and highlighting or underlining the points made when they appear. This will initiate your learning process. Without the discipline of doing this you will actually learn little, while feeling that you have learnt a lot. All the question sets in this book are like this.

Question set 6.1: API RP 1169 – responsibilities

Q1. API RP 1169: pipeline inspector responsibilities

The API 1169 pipeline inspector's main responsibility is to

(a) The pipeline manufacturer ☐
(b) The pipeline project owner or management company ☐
(c) The local jurisdiction where the pipeline will be installed ☐
(d) All of the above, depending on the contract structure ☐

Q2. API RP 1169: pipeline inspector responsibilities and relationships

While carrying out their duties, pipeline inspectors should not

(a) Comment on the pipeline contractor's work ☐
(b) Supervise the pipeline contractor's work ☐
(c) Establish a professional business relationship with sub-vendors ☐
(d) In some cases, implement cost control measures ☐

Q3. API RP 1169: pipeline hydrotesting

Which of the following specifies the role of the pipeline inspector during a hydrotest?

(a) Owner/user ☐
(b) The pipeline inspector's employer ☐
(c) API 1104 ☐
(d) OSHA guidance documents ☐

Q4. API RP 1169: emergency services responsibilities

Emergency services may be required to perform the rescue of personnel from confined spaces. API 1169 specifically states that the pipeline inspector must

(a) Hold a qualification in this activity ☐
(b) Not get involved in this activity ☐
(c) Make sure a plan is in place for these activities ☐
(d) Review and approve a plan for these activities ☐

Q5. API RP 1169: pipeline inspector role

Which of the following would not be considered as an applicable role of the pipeline inspector within the guidance of API 1169?

(a) Understand ROW arrangements and procedures ☐
(b) Re-establish any missing pipeline marker stakes ☐
(c) Be familiar with safety signs/barricades required for road and railway crossings ☐
(d) Be aware of landowner requirements e.g. where they will keep their cows during the pipeline construction programme ☐

Q6. API RP 1169: pipeline inspector involvement with planning activities

Under API 1169 which of the following statements would best reflect the activities of the pipeline inspector regarding the project planning stages of pipeline construction?

(a) Be familiar with planning software ☐
(b) Mutually plan upcoming tasks with the contractor ☐
(c) Only get involved in QA/QC planning activities ☐
(d) Not get involved, it is not the pipeline inspector's job ☐

Q7. API RP 1169: environmental contamination

Pipeline inspectors should have a basic awareness of how to identify contamination but what action should they take on discovering an environmental incident on a pipeline construction site?

(a) Immediately inform the owner/user ☐
(b) Immediately inform the regulatory authorities ☐
(c) Decide mitigation procedures ☐
(d) Implement mitigation procedures ☐

Q8. API RP 1169: pipeline inspector reporting requirement;

What should a pipeline inspector do when finding an incident of vandalism?

(a) Report it to the manufacturer ☐
(b) Report it to the police ☐
(c) Report it to the chief inspector ☐
(d) Ignore it, it's someone else's problem ☐

Q9. API RP 1169: inspector responsibilities

On a pipeline construction site, specialist inspectors assigned to blasting operations report to the;

(a) Owner/user ☐
(b) Blasting contractor ☐
(c) API 1169 pipeline inspector ☐
(d) Chief inspector ☐

Q10. API RP 1169: personnel verification

Pipeline inspectors are required to verify the qualification of personnel involved in duties that require personnel certification such as welding, blasting, operation of heavy equipment and

(a) Site security personnel ☐
(b) Corrosion control technicians ☐
(c) Site data collection personnel ☐
(d) Rights of way (ROW) application personnel ☐

Q11. API RP 1169: pipeline inspector responsibilities

During a pipeline construction API 1169 inspectors are expected to be the principal means of assuring

(a) Efficient QA/QC planning ☐
(b) Environmental compliance ☐
(c) Fitness for purpose ☐
(d) Material quality ☐

Q12. API RP 1169: pipeline inspector role

What role do pipeline inspectors have regarding other key personnel performing work on an onshore pipeline construction project?

(a) Setting personnel qualification standards ☐
(b) Organising personnel qualifications ☐
(c) Verifying personnel qualifications ☐
(d) Administering personnel qualifications ☐

Q13. API RP 1169: pipeline inspector safety responsibilities

Prior to commencing pipeline construction, API 1169 inspectors should be capable of organising and conducting

(a) Daily loss prevention system meetings ☐
(b) Daily ROW meetings ☐
(c) Daily safety meetings ☐
(d) Daily radiological protection logs for NDE examiners using radioisotopes ☐

Q14. API RP 1169: inspector responsibilities (annex A)

The API 1169 Chief Inspector shall

(a) Report directly to the owner/user and not get involved in supervising the inspection organisation ☐
(b) Be qualified as a blasting and a horizontal and directional drilling inspector ☐
(c) Be highly skilled with an in-depth knowledge in welding only ☐
(d) Be capable of implementing the project management process ☐

Q15. API RP 1169: inspector roles

The API 1169-referenced welding inspector shall be

(a) Capable of reviewing NDT technician qualifications ☐
(b) Capable of reviewing welding technician qualifications only ☐
(c) Qualified to AWS welding practice for structural steelwork ☐
(d) Also qualified as a API 1169 pipeline inspector ☐

Question set 6.2: CEPA/INGAA guide – responsibilities

Q1

If a pipeline construction contractor performs tests or measurements unassisted and without external witnessing, the pipeline construction inspector should

(a) Ask for the test to be repeated so it can be witnessed ☐
(b) Issue a 'stop work' notice until the situation is resolved ☐
(c) Issue a non-conformance notice or amend the ITP as applicable ☐
(d) Check the equipment was correctly and properly calibrated ☐

Q2

Which of the following would not be considered a role of a pipeline construction inspector?

(a) The monitoring of activities against safety requirements ☐
(b) Assisting other specialised inspectors ☐
(c) Monitoring activities against regulatory requirements ☐
(d) Planning of the contractor's progress ☐

Q3

Which of the following would not be a common safety policy/practice/procedure that would be expected to be put into place by the pipeline owner company?

(a) Hearing conservation practice ☐
(b) Environmental reporting practice ☐
(c) H_2S safety ☐
(d) Working alone policy ☐

Q4

According to the CEPA/INGAA guide, which of these is the responsibility of the inspector during pipeline construction?

(a) Appoint replacement contractors if existing ones default ☐
(b) Agree initial ROW arrangements with landowners ☐
(c) Agree any changes to ROW arrangements with landowners ☐
(d) Advance planning and organisation of all construction activities ☐

Q5

The pipeline inspector plays a critical role in managing the quality of work performed during pipeline construction. Which of these is true according to the CEPA/INGAA guide?

(a) An identified deficiency needs to be corrected ☐
(b) A non-conformance may be a one-off deficiency ☐
(c) A non-conformance is an isolated deviation from requirements ☐
(d) None of the above are true ☐

Q6

In relation to site coating during pipeline construction, a pipeline inspector

(a) Is expected to undertake all coating inspection activities ☐
(b) May not perform coating activities on their own ☐
(c) May perform spark (Holiday) testing only ☐
(d) Should also be qualified in coating application procedures ☐

Q7

When does a pipeline construction inspector have 'stop work' authority?

(a) Never, their role is limited to monitoring and reporting only ☐
(b) Only when the owner or site health & safety representative agrees ☐
(c) When there is risk of imminent danger ☐
(d) Only when a task or test has been done incorrectly ☐

Q8

During a pipeline construction project, a *push-out* is a

(a) TWS ☐
(b) Muskeg ☐
(c) Mobilisation of contractor's staff after the ROW survey is completed ☐
(d) Method of tunnelling under a watercourse ☐

Q9

The pipeline construction inspector acts as the owner company's representative and will assist with

(a) Regulatory requirements ☐
(b) The representation of INGAA during pipeline construction ☐
(c) Contract financial matters ☐
(d) Welding inspectors ☐

Q10

According to the CEPA/INGAA guide, which of these is the responsibility of the inspector during pipeline construction?

(a) Reporting people for not respecting historical resources ☐
(b) Helping people select the correct tools for jobs ☐
(c) Ensuring a drugs/alcohol policy is in place ☐
(d) None of the above are pipeline inspector responsibilities ☐

Question set 6.3: INGAA/CEPA guide – general

Q1. References

Which of the following documents should the inspector reference when looking for specific guidance for the truck transportation of line pipe?

(a) API 5L1 ☐
(b) API 5LW ☐
(c) API 5LT ☐
(d) API 1104 ☐

Q2. Code of conduct

The actual ethical conduct required from an inspector is governed by the Owner Company's code of conduct. Which of the following considerations would best describe the phrase 'behaving in an ethical manner'?

(a) Abide by confidentiality agreements ☐
(b) Manage a proactive approach to participating in the morning contractor safety meetings ☐
(c) Accept the work performed by the contractor based on their reputation ☐
(d) Complying with the relevant codes and standards that you are familiar with ☐

Q3. Owner company safety policies

One of the key roles of the inspector is to assist the Owner Company in ensuring a safe work environment for both its workers as well as the public. In support of a safe environment which one of the following safety policies/practices/procedures would not be included?

(a) Fall protection practice ☐
(b) Holiday entitlement ☐
(c) Lockout/tagout procedure ☐
(d) Job safety analysis ☐

Q4. Field bending operations

Field bending is also known as 'cold bending' because the pipe is not heated before the operation. Which of the following would not be considered as best practice by the inspector, prior to the commencement of work?

(a) Confirm appropriate instruments are available for the inspection ☐
(b) Confirm if the owner has identified ambient temperature limits ☐
(c) Confirm that the pipe was not damaged during field bending operations ☐
(d) Ensure the limitations and requirements for field bending are understood ☐

Q5. Records management

Company record keeping is vital to the long-term management of the pipeline. Which of the following would be considered a typical activity with supporting records on completion of the project?

(a) Participation in sessions in support of lessons learned ☐
(b) Obtain formal approval from the construction manager prior to commencing any extra work activities ☐
(c) Record lengths and locations of work completed on a daily basis ☐
(d) Confirm the weekly progress reports include potential cost and schedule issues ☐

Q6. Personal violations

One of the many responsibilities of the inspector is to observe and report individuals for personal violations. What would be the potential outcome for listening to the radio using headphones on the work site?

(a) Removal of worker from work site ☐
(b) Permanent removal of worker from work site ☐
(c) Construction shutdown ☐
(d) Roughhousing is not considered a personal violation ☐

Q7. Welding

The API 1169 inspector is not considered a weld inspector, although they are often asked to assist with these duties. Which of the following items would the inspector not typically include as part of his reporting requirements?

(a) Handling and the storage of welding materials ☐
(b) Hazard identification report ☐
(c) Weld mapping ☐
(d) NDE results ☐

Q8. Stockpiling and stringing

Proper storage of the pipe is considered an essential requirement when planning the stockpiling and stringing operations. What would the inspector have to specifically monitor, as part of these requirements?

(a) All equipment operators have appropriate certification ☐
(b) The correct stacking of pipe by their material composition ☐
(c) Pipe with confirmed damage is identified and stored in separate piles ☐
(d) The equipment is shut down before making repairs ☐

Q9. Ditching and excavation

Which of the following actions should the inspector take upon discovery of an historic site, as part of the ditching and excavation operations?

(a) Allow excavation of the trench to continue until told otherwise ☐
(b) The inspector should not be involved ☐
(c) Permit a deviation of up to a maximum of 10° from the planned route ☐
(d) Ensure ditching is suspended until formal approval is provided ☐

Q10. Lowering-in

As part of the lowering-in process the inspector will continually familiarise themselves with key documents including but not limited to design drawings, safety plans, permits, lift plan and

(a) Water withdrawal techniques ☐
(b) All welders' qualification records specific to the applicable WPS ☐
(c) NACE reference document SP0188 ☐
(d) Environmental protection plan ☐

Question set 6.4: CEPA/INGAA guide – inputs and outputs

Q1

Construction site clean-up is the final clearing and removal of construction materials left over the pipeline ROW. When should the final clean-up be carried out when a pipeline is constructed during the winter?

(a) Immediately after construction is completed ☐
(b) During any dry period ☐
(c) During the following spring ☐
(d) During the following winter ☐

Q2

Which of the following would a pipeline inspector not include in their reporting requirements for hydrostatic testing?

(a) A safety hazard observation report ☐
(b) Test calculations ☐
(c) Justification for the type of test performed ☐
(d) A pressure–volume curve ☐

Q3

During the stockpiling and stringing stage of a construction programme, the pipeline inspector would be expected to review reports containing

(a) Forms on load-carrying certification of trucks ☐
(b) Forms on custody transfer ☐
(c) Forms on site weather reports ☐
(d) Forms on manpower allocation ☐

Q4

During backfilling of a trench, the pipeline inspector should check that any slope-breakers are

(a) Fully removed before backfilling starts ☐
(b) Fully in place before backfilling starts ☐
(c) Removed progressively as backfilling progresses ☐
(d) Installed progressively as backfilling progresses ☐

Q5

During the survey stage of a pipeline construction programme, the pipeline inspector would be expected to

(a) Record weather conditions that caused an increase/decrease in progress ☐
(b) Ensure redline reports are complete, checked and forwarded to recipients ☐
(c) Monitor workmanship ☐
(d) Produce lowering-in reports ☐

Q6

Which of the following would the pipeline inspector be expected to report on during ditching and excavation?

(a) Weather conditions ☐
(b) Endangered species habitat ☐
(c) Location of drain tiles shown on drawings ☐
(d) Location of irrigation pipes not shown on drawings ☐

Q7

During pipeline welding, the pipeline inspector would be expected to

(a) Be aware of welding standards ☐
(b) Monitor the welding of at least 10% of joints ☐
(c) Perform weld inspection activities on their own ☐
(d) Advise the welder if they ask ☐

Q8

During the lowering-in operations, the pipeline inspector should review the information on the site lift plan, contingency plans, fire prevention plan and

(a) Specific roles of workers ☐
(b) Soil classification type ☐
(c) Ground weight-bearing capability ☐
(d) Emergency evacuation plan in case of landslide or avalanche ☐

Q9

Following the application of the coating system, the pipeline inspector report should contain details of the dry film thickness (dft), the Holiday test results and information on the

(a) Wet film thickness ☐
(b) Electrical resistance ☐
(c) Thickness profile ☐
(d) Anchor profile ☐
☐

Q10

During the survey stage of a pipeline construction programme, the pipeline inspector would be expected to review permits required for environmental, road use and crossing arrangements plus

(a) Stringing procedures ☐
(b) Reinstatement plans ☐
(c) Line lists ☐
(d) Welding procedures ☐

Q11

How quickly after lowering-in should a pipeline trench in stable soil (Type A) be backfilled?

(a) After a minimum settlement time of 24 hours ☐
(b) Not until the PI has inspected it and given approval ☐
(c) As soon as practicable ☐
(d) As soon as it has had its water-spray compaction, or it has rained ☐

Q12

A pipeline inspector's report following the installation of a Cathodic Protection (CP) system after backfilling should include details of the number of test stations installed, continuity test results, sketches of any pipeline crossings and

(a) Local soil resistance measurements at 5 km (3.1 mile) intervals ☐
(b) Distance from nearest third-party rectifiers ☐
(c) Sketches of the position of nearest electricity pylons ☐
(d) GPS locations of test stations ☐

Q13

During the clearing and grading stage of a pipeline construction programme, the pipeline inspector would not be expected to produce reports containing details of

(a) Pollution emissions from burning of riprap ☐
(b) Locations of temporary fencing ☐
(c) Soil stripping depths ☐
(d) Detailed records of blasting activity ☐

Q14

A pipeline inspector's report on site clean-up and restoration should not contain information on

(a) Potential landslip angles and calculations ☐
(b) Depth of replaced topsoil ☐
(c) Installation of additional warning signs ☐
(d) Compaction depths ☐

Q15

During ditching and excavation, the pipeline inspector would be expected to review information on

(a) Welding plan ☐
(b) Endangered species habitat ☐
(c) Topsoil segregation ☐
(d) Landowner-granted permits ☐

Q16

During pipeline welding, the pipeline inspector would be expected to provide daily reports covering the number of welds rejected and

(a) Results from hazardous atmosphere tests ☐
(b) Traceability details of weld consumables ☐
(c) Start and end locations for the shift welding crews ☐
(d) Results from a daily production weld 'test piece' sent for analysis ☐

Q17

During stockpiling and stringing, the pipeline inspector would be expected to review

(a) Material certificates ☐
(b) Traffic control plans ☐
(c) EPP plans for wetlands ☐
(d) Environmental impact assessments ☐

Q18

During the clearing and grading stage of a pipeline construction programme, the pipeline inspector would be expected to review contracts for clearing, grading, timber salvage and

(a) Inspection of excavation and lifting equipment ☐
(b) Trench excavation ☐
(c) Relocation of livestock ☐
(d) Road use ☐

Q19

During the field bending stage of a pipeline construction programme, the pipeline inspector would be expected to produce reports containing information on the number/type of bends made, as-built information of the bends and

(a) GPS location data ☐
(b) Labour hours worked ☐
(c) Trench depth and width ☐
(d) Certification of machine operators ☐

Q20

During the survey stage of a pipeline construction programme, the pipeline inspector would not be expected to review welding procedures and

(a) Line lists ☐
(b) Third party crossing permits ☐
(c) Reinstatement plans ☐
(d) Environmental permits ☐

Question set 6.5: CEPA/INGAA guide – monitoring activities

Q1

During the installation of a cathodic protection (CP) system for a buried pipeline, the pipeline inspector should ensure that test lead conduits are located

(a) Any location is acceptable
(b) To the left of centreline of pipe when facing downstream
(c) To the right of centreline of pipe when facing downstream
(d) Above the pipe, so it is more easily accessible

Q2

During field bending, any pipe that does not meet the wrinkle specification shall be marked and

(a) Heated before attempting to re-bend the piping
(b) Hydrotested to check its integrity
(c) Removed from the ROW
(d) Have radial grooves ground in the weld bevels to denote it cannot be used

Q3

What action should be taken if sinkholes are found along the ditch line during a backfill operation?

(a) Work shall continue as normal as this is not unusual
(b) Work shall stop for consultation
(c) They shall be filled in immediately with backfill
(d) They shall be filled in immediately with topsoil

Q4

The pipeline inspector should continuously monitor the lifting and lowering-in operation for safety and to ensure no damage occurs to the pipeline and associated coating. Which of the following statements would you consider as incorrect following the lowering-in of a coated pipeline into a trench?

(a) Sag bends should be firmly supported ☐
(b) Ensure that the pipe is in the centre of the trench ☐
(c) Ensure that the coated pipe is never dragged along the base of the trench ☐
(d) Side bends should be supported by the trench walls to reduce stress ☐

Q5

When side boom machine operators are traversing under powerlines they should

(a) Lower the boom to ground level ☐
(b) Lower the boom to maximum 6 feet above ground level ☐
(c) Move at maximum speed of 5 mph ☐
(d) Use spotters ☐

Q6

Where sand is used for padding before backfilling a trench the sand should be

(a) Dry ☐
(b) Dampened ☐
(c) Mixed with rocks and hard materials for added strength ☐
(d) Mixed with top soil ☐

Q7

Which of the following would be considered as the main focus of a pipeline inspector when witnessing/monitoring the lowering-in of the pipeline into a trench?

(a) Coating integrity ☐
(b) Safety of the trench shoring arrangements ☐
(c) The condition of the trench bottom ☐
(d) All of the above ☐

Q8

Which of the following is a prerequisite prior to the commencement of ditching operations?

(a) Trench breakers shall be in place ☐
(b) All personnel shall be removed from a 50 feet area either side of the ROW ☐
(c) All personnel shall be removed from a 20 feet area either side of the ROW ☐
(d) An 811 certificate needs to be in place ☐

Q9

During installation of a pipeline CP system, the pipeline inspector should

(a) Test leads again after backfilling ☐
(b) Validate the CP laboratory test readings ☐
(c) Report any CP test stations that are accessible from nearby roads ☐
(d) Ensure junction box connections are made after burying the lead wires ☐

Q10

What should be used to lift pipe sections from transport trucks during a stringing/stockpiling operation?

(a) A fork lift truck fitted with recessed wooden bearers ☐
(b) Braided slings or fibre straps (2 minimum) ☐
(c) Any arrangement excluding fibre straps ☐
(d) Metal end hooks ☐

Q11

Which of the following methods would not be permitted by the CEPA/INGAA guide for the transport and handling of piping during a stringing and stockpiling operation?

(a) Fibre straps ☐
(b) Reinforced fibre straps ☐
(c) Chains or metal straps ☐
(d) Slings 'double-passed' around the pipe circumference ☐

Q12

When inspecting pipe sections during a stringing and stockpiling operation, the pipeline inspector should

(a) Make eye contact with the lifting equipment operator ☐
(b) Ensure lifting operations are stopped when it is raining ☐
(c) Not do inspections when lifting equipment is in operation ☐
(d) Ensure that the pipe section is laid on the ground ☐

Q13

Before lowering-in of a coated pipeline into a trench the trench should

(a) Have any permanent trench-breakers installed ☐
(b) Be drained ☐
(c) Contain wooden bearers so the pipe can be adjusted for longitudinal position ☐
(d) Contain no sandbags ☐

Q14

When inspecting coated pipe sections during a stringing and stockpiling operation, the pipeline inspector should expect to see all pipe sections marked with size, wall thickness, date of coating and

(a) Flange rating and/or hydrotest pressure ☐
(b) Coating vendor ☐
(c) NDE technician's stamp ☐
(d) QA plan reference ☐

Q15

Pipeline weld radiography is performed

(a) After coating and hydrotest ☐
(b) Before coating and hydrotest ☐
(c) On weld edges before welding ☐
(d) Before coating and after hydrotest ☐

Chapter 7

Line pipe materials: API 5L

7.1 API 5L line pipe

The lengths of pipe spools used to make up a pipeline are normally referred to by the generic term *line pipe*. The most common steel grade used is API 5L, equivalent to the European designation ISO 3183. This is a basic carbon–manganese (C–Mn) steel dosed with a small proportion of trace elements such as sulphur, silicon, vanadium, titanium and niobium to improve its properties. API 5L comes in eleven basic grades numbered A25, A, B, X42, X46, X52, X56, X60, X65, X70 and X80. These are in increasing order of specified minimum yield strength (SMYS), with the last two digits being the SMYS in ksi ($\times 1000$ psi). Hence X52 has a SMYS of 52,000 psi.

API 5L product specification levels (PSLs)

Most of the API 5L grades can be ordered in one of two product specification levels (PSL 1 or PSL 2). PSL 1 represents standard quality and is generally used when supply quality is assumed to be reliable and when no special low-temperature ($<0°C$) properties are required for the material. Increased quality level PSL 2 specifies tighter control of trace element levels and improved testing requirements. Figure 7.1 shows the details for API 5L grade X52. The principles are much the same for the other API 5L grades.

Problems with API 5L steels normally occur with the high-strength grades X52 and above. With the PSL 1 class there is no maximum limit placed on strength values. If material is supplied with strength levels in excess of those allowed by the PSL 2 class, then they become difficult to weld without cracking, particularly in cold weather. Routine Charpy impact tests are not required for PSL 1 so any material that is excessively hard and brittle (as high-strength steels can be) will escape

FIG 7.1
Properties of a typical API 5L line pipe material

	API 5L grade X52 line pipe	
	PSL 1 Standard quality	PSL 2 Special quality
Carbon	0.024%	0.28%
Manganese	1.4%	1.4%
Phosphorus	0.025%	0.03%
Sulphur	0.015%	0.03%
Silicon	0.45%	—
Vanadium	0.1%	Sum of niobium +
Niobium	0.05%	vanadium
Titanium	0.04%	+ titanium ≤0.06%
Specified minimum yield strength (SMYS)	52 ksi min	52 ksi min 76.9 ksi max
Ultimate tensile strength	66.7 ksi min	66.7 ksi min 110.2 ksi max
Charpy (input tests)	None required	Mandatory
NDE (seamless)	Only at purchaser's request	Mandatory
Traceability	Until tests are passed	Mandatory to completion of delivery
Hydrostatic test	Mandatory	Mandatory

detection, with problems arising at the welding stage or later in cold weather service.

Pipe diameter and wall thickness

Line pipe diameter varies from 4 inches to up to 42 inches – the most common size used on long-distance overland pipelines. Larger sizes are available (up to 72 inches) but are limited by the capability of manufacturers. Compared with other types of pressure equipment, line pipe is quite thin-walled, ranging from $\frac{1}{2}$ inch to 2 inches for higher pressure applications. Most, however, are in the $\frac{1}{2}$ to 1 inch range.

Pipe wall thickness is determined by either of the two US systems of pipe sizing. The older system divides pipe sizes into standard (STD), extra strong (XS) and extra extra strong (XXS). Alternatively, the 'schedule number' system lists schedule sizes from Sch 10 up to Sch 160. Figure 7.2 shows a typical example for 20 inch and 24 inch sizes. Above Sch 100 the pipe wall thickness becomes quite thick, requiring much more attention to be given to welding and post-weld heat treatment (PWHT) techniques.

FIG 7.2
Illustration of line pipe schedule sizes

Nominal pipe size (inches)	Sch	Outside diameter (inches)	Inside diameter (inches)	Wall thickness (inches)	Weight (lb/ft)
20	100	20.000	17.438	1.281	256.1
20	120	20.000	17.000	1.500	296.4
20	140	20.000	16.500	1.750	341.10
20	160	20.000	16.062	1.969	379.2
24	10	24.000	23.500	0.250	63.41
24	20	24.000	23.250	0.375	94.62
24	30	24.000	22.875	0.562	140.7
24	STD	24.000	23.250	0.375	94.62
24	40	24.000	22.625	0.688	171.3
24	60	24.000	22.062	0.969	238.4
24	XS	24.000	23.000	0.500	125.5
24	80	24.000	21.562	1.219	296.6
24	100	24.000	20.938	1.531	367.4
24	120	24.000	20.375	1.812	429.4
24	140	24.000	19.875	2.062	483.1
24	160	24.000	19.312	2.344	542.1

7.2 Manufacture

Seamless or welded?

API 5L line pipe can be made in a manufacturing shop in either seamless or longitudinally electrical resistance welded (ERW) form. Both types still need to be circumferentially welded during assembly at the construction site and the manufacturing process has no effect on this. The choice of which is used on a particular project comes down mainly to cost and availability. ERW pipe has a better surface finish and is more expensive. Seamless pipe is sometimes preferred to ERW in small sizes (> Sch 100) as it doesn't present the problem of a small-diameter weld to radiograph. For design purposes (ASME B31.8) there is no difference in design factor, and hence design wall thickness, between ERW and seamless, so the weld is not seen as an inherent source of defects as long as it is subject to the correct NDE. Worldwide, the trend is towards the use of ERW line pipe in large diameters, driven perhaps by better quality of welding than in previous years when weld failures in service were more common than they are now.

FIG 7.3
How ERW line pipe is made

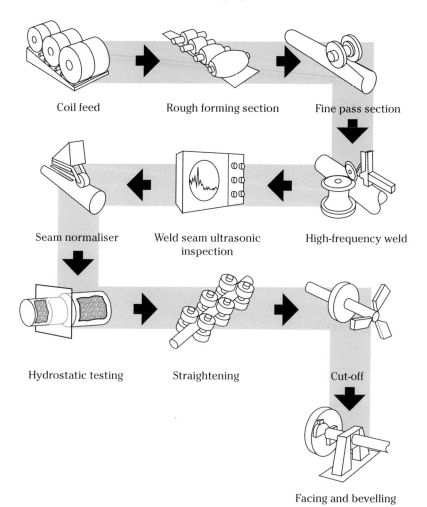

How ERW pipe is made

Figure 7.3 shows the steps in manufacturing ERW line pipe. The sheet steel coil is unrolled and formed roughly into a circular shape using 'first form' rollers. The longitudinal joint is then closed up using rollers with reverse taper profiles and then automatically welded using a high-frequency resistance weld. The weld is then heat treated and cooled

FIG 7.4
How seamless line pipe is made

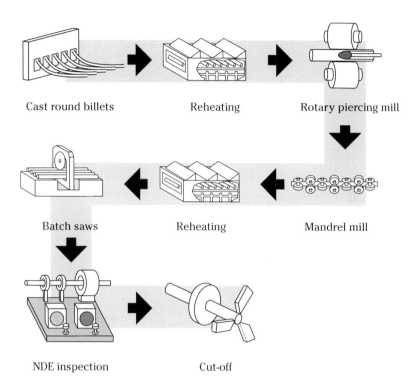

(normalised) and ultrasonically tested for defects. All of these are automated, continuous production processes. To finish the pipe lengths they are hydrostatically tested, straightened between a bank of vertical rollers and cut to length. Finally, the cut faces are faced and bevelled to the joint shape required for welding.

How seamless pipe is made

There are several ways of doing this. The pipe starts off as a hot-forged billet, of circular cross-section. After reheating, it is pierced down its centreline on a rotary piercing mill then progressively stretched out using various roller/mandrel arrangements to a semi-finished form. Figure 7.4 shows the procedure. The extreme amount of working required means that it has to be reheated and subjected to further stretching and forming to produce the final required shape. This is followed by automated NDE and facing/bevelling in a similar way to ERW pipe.

7.3 Inspection issues

Traceability

The continuous production nature of line pipe manufacture means that traceability is not such a problem as it can be for pressure vessels assembled from components with multiple origins. Although API 5L PSL 1 has fairly relaxed certification requirements, line pipe from most manufacturers is automatically marked (by stamping or etching) as it is produced in the mill, so mix-ups are rare. Any that do occur invariably happen in the stockholding yard when material is being stored for different construction projects. As with any product, there is 'rogue material' available on world markets so there is still a role for the site pipeline inspector (PI) in checking and validation. PSL 2 has mandatory material certificate requirements, although these can be issued and validated by the material manufacturers themselves, e.g. a '3.1B class' certificate.

Surface condition

API 5L contains requirements for the surface condition of the finished pipe. This covers features such as

- weld undercut (for the longitudinal shop weld in seamed pipe)
- arc burns
- laminations
- hard spots.

Dimensional tolerances

Dimensional accuracy is important both for the integrity of the pipe under pressure and its influence on the circumferential welds that will be used to join the pipe sections on the construction site. Important controlled dimensions are

- tolerances on outer diameter and wall thickness
- out-of-roundness (OOR)
- peaking at the longitudinal weld
- height of longitudinal weld bead

- alignment
- straightness.

Markings

Chapter 11 of API 5L gives extensive details on how pipe spools should be marked before despatch, including stamping, etching and colour-coding. These are points that should be checked by the PI as the spools arrive on site.

Chapter 8

The pipeline construction codes: ASME B31.4 and ASME B31.8

The two construction codes ASME B31.4 *Pipeline Transportation Systems for Liquids and Slurries* and ASME B31.8 *Gas Transmission and Distribution Piping Systems* are the most common codes used worldwide in the pipeline industry. They have been refined and amended over many years and are adopted by most of the US pipeline industry. Codes with different names but with similar principles and content have been adopted for use in Canada, the Middle East and Asia.

The latest version of the API 1169 BoK published by API shows that both B31.4 and B31.8 have been officially *removed* from the BoK. They were previously listed in the open-book question section. Their new status is 'for guidance only', inferring that they will not be used as a direct source of exam questions. This is not as big a change as it appears: many of the B31.4/B31.8 exam-question-style requirements are given in the two sets of federal regulations 49 CFR 192 and 49 CFR 195 (see Chapter 12 of this book), simply repeated word-for-word. Hence they are effectively still in the BoK, just in a different document.

Notwithstanding this, the ASME B31.4/B31.8 codes contain useful guidance points to add to the knowledge of the pipeline inspector (PI). The general knowledge points still make valid closed-book exam questions on the basis that the *principles* of ASME construction codes are essential knowledge for plant inspectors of all types. To explain, let's take a short step backwards and see what ASME construction codes actually *are*.

8.1 ASME construction codes – what are they?

Let's start with what they're *not*. No construction code purports to provide the single best way to design and manufacture a pipeline, vessel, valve or anything else. This is because there will always be different ways to do things, each with its positive and negative points. Each of these depends on individual differences and circumstances, known and unknown, that inhabit the world of engineering. Here's what a construction code *does* provide.

- A variety of practical solutions for design, manufacture, testing and so on that have been *shown to work*.
- Some *mandatory* requirements that experience has shown are necessary for the safety and integrity of an item.
- A few *prohibited* features and activities that are considered a real risk to safety and integrity, and therefore must *not* be used.
- Personnel roles, responsibilities, activities and qualifications considered necessary to ensure the design, manufacture, testing and certification of a component is carried out in a competent and organised way.
- Provide cross-referenced links to other codes and standards (both technical and quality-based) needed to complete the picture.

Taken together, these code 'outputs' apply to engineering equipment codes in general, and static pressure equipment in particular. Rotating equipment, electrical and other codes follow a similar model, but with different technical content. You may have noticed that none of these outputs claim that a published code pretends to be a document offering technical *excellence*. Apart from the fact that there can be no simply-agreed definition of what excellence actually is, any solution to an engineering problem that offered excellence would, almost by definition, be an over-engineered and over-expensive solution to what might be a fairly straightforward engineering requirement. This is exactly the situation with most items of static pressure equipment; their function is straightforward and there is little that is fundamentally new in the way of materials, design or fabrication that warrants a search for new and finely tuned innovative solutions. New and expensive solutions to already adequately solved problems owe more to vanity than engineering excellence.

ASME code contents

Think for a moment about what you would need to know if you wanted to design and construct a pressure equipment item from scratch.

First, you would need to be sure that the code you had been given to follow was indeed applicable to the item you wanted to make, so you'd need to check its *scope*. Next would come the *materials* you needed to make it from and how to calculate *design* thicknesses and flexibility (if it was long and thin enough to bend). Materials don't come pre-prepared in the correct shape so you'd need a guide on *fabricating and welding*. To check the quality of your work, *NDE, inspection and testing* are then required. Finally, to keep quality assurance (QA), quality control (QC) and other paper-shufflers all happy, suitable *documentation and certification* would be nice to have.

There, in essence, is the content of a construction code. It's a fairly simple formula. The only thing that makes the real-world documents look long and complicated is their need to take into account all the possible variations and options to which the code may need to apply (different materials, risk classes of pipework, forged components, cast components, different types of welding etc.). These can outnumber the base code content by a factor of four, five or more, giving the need for sprawling series of annexes and appendices, and a spider's web of cross-references to link the whole lot together. Once you recognise this format, construction codes do not really seem that difficult at all.

Now let's try to apply this approach to ASME B31.8 and see how it works.

8.2 ASME B31.8: what's the scope?

There it is: section 802.1, right at the beginning, limiting it to liquefied petroleum gas (LPG) transmission pipelines, setting out the upper and lower temperature limits and excluding all the equipment it doesn't apply to: piping inside refinery sites, vessels, heat exchangers and similar. It has to do this as these things fall within the scope of other construction codes, and duplication would be confusing.

Why all the definitions in section 803?

ASME codes love their terms and definitions section. They don't make particularly exciting sequential reading because they are there to *refer back* to if you need to clarify the meaning of something in the text. That's also the way they are written too, with the code text completed first. ASME B31.8 is unusual in the way that the terms and definitions are divided up into six or seven different technical categories, each in its own alphabetical order.

Anything special about acceptable materials?

Chapter I gives a list of acceptable new material specifications (API 5L, ASTM A106 etc.). A notable point is the provision for use of previously-used pipeline sections (811.3) as long as they comply with listed specifications and/or mechanical tests.

Chapter II: welding requirements

Welding procedures, and welders themselves, have to be qualified – a fairly common requirement. ASME pipeline codes are a little unusual on allowing a choice of welder qualification routes for pipeline operating under a low hoop stress of <20% specified minimum yield strength (SMYS). Above this, qualification to ASME IX for new construction, or API 1104 for in-service welding, is required.

Welding inspection requirements are listed in section 826 of chapter II. Welds require visual examination and, in addition, field butt welds between pipeline sections have to be inspected by NDE. The extent is

- Location class 1: 10%
- Location class 2: 15%
- Location class 3: 40%
- Location class 4: 75%.

Chapter III: design

Design features controlled by the technical requirements of ASME B31.8 chapter III are the same as for all other pipework/pipeline construction codes, i.e.

- pressure design (wall thickness)
- closures (flanges and blanks)
- reinforcement of nozzles and branch connection
- bending flexibility
- reaction at supports.

Chapter IV: installation and testing

Unlike pressure vessel codes, pipeline codes include requirements and restrictions for the *installation* of a pipe in its trench. Four location classes are defined (see Figure 8.1) based on the consequence of a potential failure to nearby people and property. Minimum depths of backfill cover are given in a table. This is the same requirement as that in federal regulation 49 CFR 192.

FIG 8.1
The gas pipeline location classes (ASME B31.8)

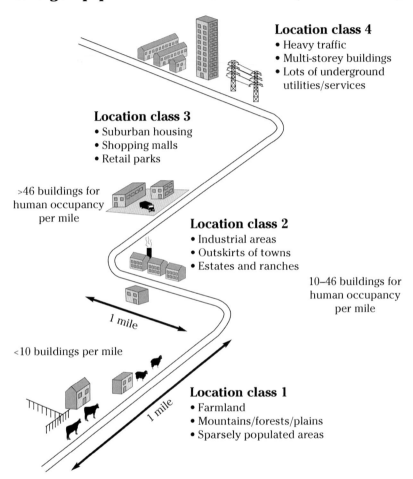

The location class decides what design (efficiency) factor is used when calculating the pipeline wall thickness (between 0.4 and 0.8) and also its hydrostatic test pressure (Table 841.3.2 (1))
For a full description see ASME B31.8 (840.2.2) and Table 841.1.6 (2))

Under ASME codes, pressure testing may be performed using water (hydrostatic) or inert gas (pneumatic). Test pressures vary with location class, to reflect the perceived risk level involved.

Finally, ASME B31.8 section 861: corrosion control

This is almost unique in the world of construction codes. Most end their scope of influence at the point where the product leaves the manufacturing works accompanied by all its control documentation and test results. The main aspects are the pipeline external surface preparation coating and cathodic protection (CP). There's not a lot of technical detail contained in ASME B31.8 (or B31.4) about these subjects, just statements of general requirements, so cross-references to other codes and published standards are used to fill in the gaps.

8.3 The liquid pipelines construction code: ASME B31.4

ASME B31.4 *Pipeline Transportation Systems for Liquids and Slurries* is the ASME construction code for pipelines carrying fluids that exist as stable liquids rather than liquefied gas. The main liquids quoted in the scope are hydrocarbons (oil products), LPG, anhydrous ammonia, alcohols and slurries. The content covered by the API 1169 BoK is much the same as we saw for ASME B31.8 for gas pipelines (i.e. scope, materials, welding/construction inspection and testing). However, unlike B31.8, nothing is included on installation in the trench. Although the scope covers liquids that may be chemically hazardous, the assumption is that liquids do not carry the same level of explosive risk as liquefied gases. We can therefore fairly consider ASME B31.4 as a being a lower risk pipeline code, compared with ASME B31.8.

How to interpret the ASME B31.4 'code scope' diagram figure 400.1.1-1

ASME pipework and pipeline construction codes generally contain a code scope diagram to illustrate which parts of a system fall within the scope of the code in question and which fall outside, to be covered by other codes. These can look confusing at first viewing but are simple enough once you actually read them and understand their principles. Figure 8.2 shows a reproduction of the scope diagram figure 440.1.1-1 in ASME B31.4. It works like this.

- System flow starts at the top of the diagram at the production field and runs to the bottom at the product unloading terminal. The product passes, or is diverted, through various refining and storage facilities along the way.
- Physical boundaries (termed 'plot limits') of the various plant facilities are indicated by the *dashed lines*, either light or heavy weight as per the key at the bottom of the figure.
- Piping, shown as the thick black line (look at the key), falls *within the scope* of ASME B31.4. The important part from this code is how the B31.4 scope continues across and into the plant areas, rather than stopping at the physical plot limit or property boundary (the fence).
- Once the thick black line is inside the plot limits then what happens to it next shows how the ASME B31.4 code coverage continues. If it is shown ending with a straight perpendicular 'T' at the end of the line, then B31.4 coverage ends (and starts) somewhere within the plot limits – perhaps it joins to a pressure vessel (ASME VIII) or enters a liquid storage tank (API 650). If it ends with a squiggle, then B31.4 continues to apply to the system as it passes around inside the plot limits, before exiting on the other side.

This same pattern of line style and weight is followed in most ASME code scope diagrams. That's all there is to it. It also makes good open-book exam questions.

ASME B31.4 inspection and testing

ASME B31.4 mentions the role of a PI in clause 436.2. There is no specific reference to API RP 1169 because B31.4 has been around for a lot longer and the documents are from different organisations, which never helps. The roles and competence of the PI are listed in the same way they are in both RP 1169 and the INGAA/CEPA guide, starting from the activities of RoW surveying, clearing and grading, through to trenching, lowering-in, backfilling and pressure testing. There is a bit of detail given, but you have to go to the CEPA/INGAA guide for full lists of requirements.

Pressure testing (437.1.4) specifies a full 4-hour hydrostatic test at minimum $1.25\times$ MAOP (maximum allowable operating pressure) for pipelines that will be operating at a hoop stress $> 20\%$ SMYS. This may be replaced by a 1-hour leak test, either hydrostatic (at the same $1.25\times$ MAOP) or alternatively by pneumatic testing at the much-reduced pressure of 100 psi, for safety reasons. In practice, pneumatic leak tests are much more sensitive at finding leaks than hydrostatic tests, but

FIG 8.2
Diagram showing scope of ASME B31.4
excluding carbon dioxide pipeline systems (See fig. 400.1.1-2)

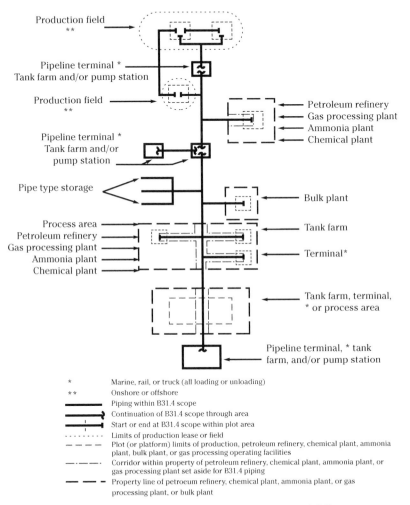

Reproduced courtesy of American Society of Mechanical Engineers (ASME)

necessary safety precautions must be taken (see API RP 1110 and INGAA CS-S-9).

For the rest of ASME B31.4, the design and fabrication and welding requirements follow the same general pattern as ASME B31.8. We can maybe anticipate that B31.8 would be a preferable source of exam questions to B31.4, purely because gas pipelines represent a higher risk, so the PI's knowledge of this code is that much more important.

Training verification – watch out for the hidden detail in ASME B31.8

Depending on who you ask, you will hear opposing views that ASME codes are either a shining example of pragmatism or a mixture of lowest-common-denominator technical requirements, with bits of devil hidden in the detail. Whichever is correct you can certainly find lots of hidden details if you look. Most of these involve detailed design issues such as factors in equations and have no relevance to the role of the PI. Occasionally, however, one pops up that does fit within the PI's remit, influencing the monitoring/verifying/recording role of the PI as set out in RP 1169 and the CEPA/INGAA guide. One such example can be found hiding away in ASME B31.8 section 807 *Training and qualification of personnel.*

Have a brief look at ASME B31.8 section 807. It sets the need for training programmes to be in place for those personnel working for a pipeline operating company that need them. A short step takes you to the need to identify which personnel roles actually *need* a training/qualification structure in place (i.e. those that involve activities that could impact the safety or integrity of the pipeline). Fair enough – RP 1169 clearly charges the PI with the job to verify that all such training activities and qualifications are correctly in place: report these to the owner/user and all will be well.

Now for the hidden detail.

Take a look at Figure 8.3. This has been extracted from the detail of section 807.1. What at first glance appeared to be the innocuous term *training programmes* has taken on a new complexity when we see what B31.8 says this 'programme' must actually consist of. Here are some of the more awkward requirements, picked out so you can see them.

- Personnel must be trained to recognise *abnormal operating conditions*, as defined in ASME B31Q. That means they need training about it.
- The 'training programme' that the PI is to verify must contain steps

FIG 8.3
Hidden detail
– The B31.8 (807.1) training programme 'loop' –

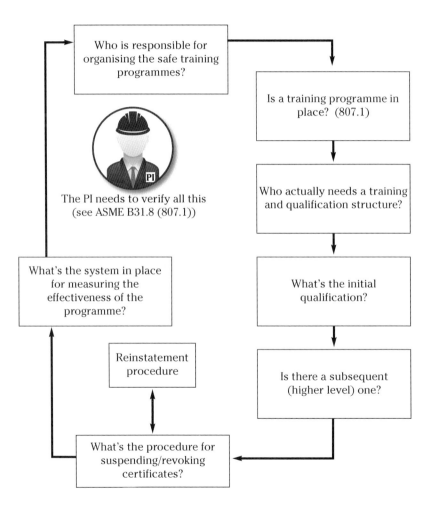

for evaluating criteria used to determine *suspensions (or reinstatement)* of qualifications.
- There must be a process in place to periodically evaluate the *effectiveness* of the site qualification programme. There it is in 807.1 (f).

So what's the problem?

The problem is that, given the fluid nature of personnel and contractors present on a long pipeline project (some go on for years), full compliance with the 'assurance loop' shown in Figure 8.3 is often not easy to achieve. Long projects with stretched logistics have a tendency to cause loops to break down as people move, change roles and day-to-day project problems occur and have to be solved first. The result is that this is a common area of non-compliance and it's the PI's job to find and report it. This, of course, is good rather than bad news. It is the PI's job to find non-compliances where they exist, not just look for evidence that everything is all right. Non-compliances, as we now know, can sometimes reside in the detail. Now read 807.1 through again, and this time you will see it.

8.4 API 1169 exam questions

As mentioned earlier, direct-quote exam questions sourced from ASME B31.4 or ASME B31.8 are unlikely because the codes are no longer included in the official exam BoK. Remember, however, that the principles of construction code content are a valid knowledge area for PIs and that key bits of information in ASME B31.4/31.8 also appear in federal regulations 49 CFR 192 and 49 CFR 195, which are included in the open-book question section of the BoK.

Chapter 9

The welding code: API 1104

9.1 The role of welding codes

API 1104 *Welding of Pipelines and Related Facilities* (and its Canadian equivalent CSA Z662-15) are the main welding documents included in the API 1169 Code Effectivity List. API 1104 is well established in the pipeline industry throughout the world. It was first published in 1953 and has developed fairly slowly, retaining much of its content over 20+ editions since.

As we saw in Chapters 4 and 5, the role of the pipeline inspector (PI) is a wide one, incorporating project, safety and environmental responsibilities well outside the purely technical duties of the traditional manufacturing source inspector (SI). For this reason, the PI's involvement in pipeline weld inspection is a small one – it is not the major technical knowledge area that it is in other API ICP examinations. You can see this by looking at the API 1169 BoK: surveillance and inspection of welding questions form only about one tenth of the 50% of exam questions drawn from the pipeline construction activities part of the BoK. That's 5% of the total questions. Add another 5% or so for more general questions related to qualifications, certifications and similar and you still have only 10%.

In recognition of the limited role of the PI in pipeline weld inspection, annex D of RP 1169 specifies the duties of a separate welding inspector (WI). This is a widely recognised inspector role for all static fabricated equipment – WIs have normally completed training and certification to ASME, AWS, CWI (certified welding inspector) or API 1104 standards. It is a full-time role in a pipeline construction team, normally employed by the pipeline main contractor. Monitoring of the WI is the job of the PI, alongside all the other monitoring duties set out in RP 1169.

9.2 So what about API 1104 itself?

API 1104 is perhaps best described as a utilitarian code. Think of it as a specification for technical *acceptability* rather than technical excellence and you won't be far wrong. Its use is limited to butt jointing of sections of pipeline into a continuous length, with the occasional fillet-welded reinforcing sleeve or attachment for good measure. Pipeline welding is a repetitive process that is well suited to modern automatic/mechanised welding techniques, particularly on larger diameter pipelines with easy internal access. API 1104 has adapted (slowly) to these developments – there are a couple of later sections on automatic techniques, but most of the document sticks to descriptions relating more to traditional manual or semi-automatic techniques. Note also that it is only these techniques that are covered in the API 1169 BoK.

API 1104 is (by accident or by design... who knows?) a fairly flexible code. It differs from most pressure equipment codes in that it allows significant welding-related indications such as lack of penetration or lack of fusion to remain in a completed weld without being classed as an 'out-of-code' defect. It uses welding and indication definitions based on the AWS 3.0 *Glossary of terms* document, different to those used in static pressure equipment codes such as ASME VIII, the ASME B31 pipework series, ASME V and ASME IX. In many ways, API 1104 resembles a structural steelwork code rather than one for pressure equipment. The principles may be the same, with similar control over the welding procedure, but the defect acceptance is more flexible.

API 1104 and the API 1169 exam BoK

The API 1169 BoK only covers the following sections of API 1104.

Section 3: Terms and definitions
Section 4: Specifications (not much in here)
Section 5: Qualification of weld procedures
Section 6: Qualification of welders
Section 7: Design and preparation of weld joints
Section 8: Inspection and testing of the completed welds
Section 9: NDE acceptance criteria
Section 10: Repair of weld defects
Section 11: NDE procedures (i.e. the techniques)

Section 13, about automatic welding, and the three annexes are not included.

Section 3: Terms and definitions

In common with other API codes, API 1104 (section 3.1) confirms the definition of the terms *indication, imperfection* and *defect* (see Figure 9.1). Apart from these, most of the terms are fairly straightforward. Section 3.2 contains a list of specific acronyms for indications found by NDE. These are specific to API 1104 and need-to-know information for exam questions. Some of them are a bit unusual. Here they are.

- AI: Accumulation of imperfections
- BT: Burn through
- ESI: Elongated slag inclusion
- IC: Internal concavity
- ICP: Inadequate cross-penetration
- IF: Incomplete fusion
- IFD: Incomplete fusion due to cold lap
- IP: Inadequate penetration 'without high–low'
- IPD: Inadequate penetration 'due to high–low'
- ISI: Isolated slag inclusion
- IU: Undercut adjacent to root pass
- VC: Volumetric cluster
- VI: Volumetric individual

Acceptance levels and illustrations for many of these indications are provided in API 1104 section 9.

Section 4: Specifications

This is a short section related to the specification of equipment, filler metals, fluxes and gases used for the welding process. A specific requirement is that all filler metal and fluxes must conform to one of a list of nine specific AWS standards – not all welding codes do this. The list of material types (between AWS A5.1 and AWS A5.29) allows for all the different steels used in pipeline manufacture.

Section 5: Qualification of weld procedures

There is nothing innovative about this section – it just outlines the requirement for qualifying weld procedure specifications (WPSs) using procedure qualification records (PQRs) and the qualification of the welder to match. Figure 9.2 shows the idea. Sub-sections contain lists of the essential variables following the same well-established principles as ASME IX. Overall, these are probably too detailed to be used for

FIG 9.1
Key definitions

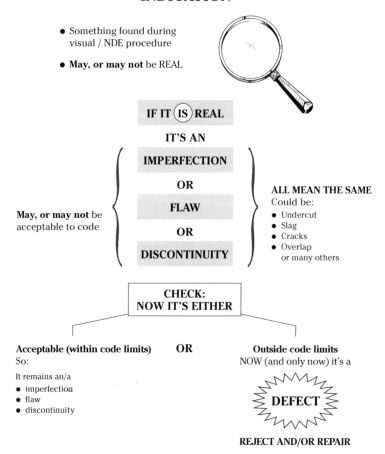

FIG 9.2
Welding: controlling documentation

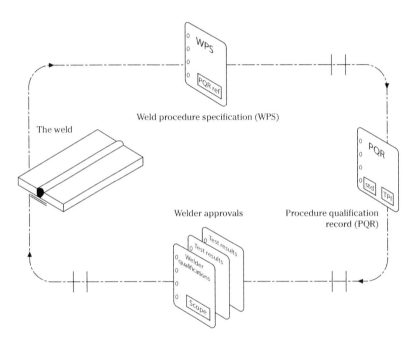

closed-book API 1169 examination questions. They are more suited to API SIFE or API 510/570-style exams where candidates' welding knowledge makes up a larger part of the scope of the question bank.

The types of mechanical tests done when qualifying a WPS are considered general welding knowledge valid for the exam. Questions are likely to be straightforward, about the types of test, rather than their technical detail. The main tests are

- tensile strength test (uses a flat specimen)
- nick-break test
- root bend test
- face bend test
- side bend test.

FIG 9.3
Pipeline welding
– Destructive test specimen for PQRs –

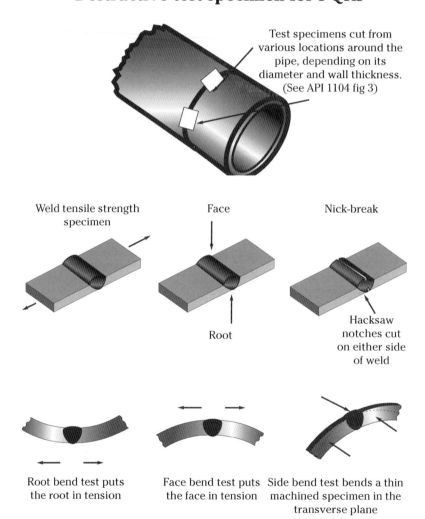

There is a lot of detail in API 1104 section 5 (see table 2) about how many of each test are required for PQR testing (see Figure 9.3 for the basic detail and where the specimens are cut from). This is a bit too detailed for closed-book examination questions. Note the general principle, however, that the greater the pipeline diameter and wall thickness, the more test pieces (particularly for the tensile test) are required. This is due to the greater thickness being more likely to have problems with excessive welding heat input, internal defects and brittleness.

Section 6: Qualification of welders

Again, this resembles a simplified version of the requirements of ASME IX, with a welder preparing a test piece to the WPS configuration then submitting the piece for visual, non-destructive and destructive tests.

Note the role of the WI
Arguably the PI is only required to provide high-level monitoring of the process of qualifying welders to the requirements of API 1104. Most of it is done by the WI. This is confirmed in annex D.2.3 of RP 1169, which states that the WI must be capable of monitoring and assessing all the various tests and (in D.2.a) the documentation and records that go with them.

Section 7: Design and preparation of weld joints

Section 7 covers inspection of the *weld joint*, the name given to the set-up before welding takes place. Emphasis is on correct alignment of the wall thickness to be joined (maximum allowable offset of $1/8$ inch (3 mm)) and methods of achieving this using alignment clamps, depending on whether positional or roll welding is to be used.

Pipeline joining welds are a fairly simple arrangement. In thinner wall sizes a single bevel joint is used. Figure 9.4 shows a typical arrangement. Single-sided welds are welded from the outside of the pipe. Root gap and land dimensions are $1/16$ inch with an included bevel angle of $60°$. The joint is aligned by attaching an alignment clamp to both sides of the joint and then the root pass added, normally using a downwards welding technique from either side of the vertical axis of the pipe. The root pass is ground back from the inside to eliminate any slag that has accumulated at either side of the weld convex profile.

A single-run hot-pass weld is then added immediately afterwards,

FIG 9.4
Pipeline welding

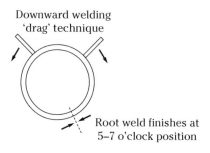

Downward welding 'drag' technique

Root weld finishes at 5–7 o'clock position

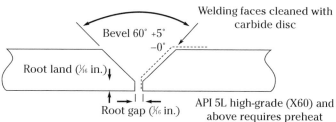

Bevel 60° +5° −0°

Welding faces cleaned with carbide disc

Root land (1/16 in.)

Root gap (1/16 in.)

API 5L high-grade (X60) and above requires preheat

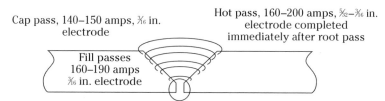

Cap pass, 140–150 amps, 3/16 in. electrode

Hot pass, 160–200 amps, 5/32–3/16 in. electrode completed immediately after root pass

Fill passes 160–190 amps 3/16 in. electrode

Root run, 125–165 amps, 5/32–3/16 in. electrode E-XX10. Grind back to eliminate excess convexity, then remove the alignment clamp

- This shows a typical manual SMAW technique. Typical electrodes are E6101, E7010 or E8010
- The purpose of the hot pass is to burn out lines of slag (wagon tracks) at the edges of the convex root pass weld
- Higher current used for downward welding enables quick welding
- Always consult the WPS/PQR approval for use on the site

followed by full passes of increasing width. Finally, the cap pass completes the weld. Electrode sizes and types can vary with WPS arrangement; Figure 9.4 shows a typical example.

Stronger X-grade material
Lower API 5L 'X-grades' of pipeline material can normally be welded without preheat. However, if the stronger grades (X60 and above) are used, then preheat becomes necessary, plus a more refined WPS to minimise the increased chance of weld cracking. This is always the risk with using higher-strength API 5L alloys.

Section 8: Inspection and testing of the completed welds

From the PI's perspective, the main emphasis here is the verification of qualifications and certification of weld inspection and NDE personnel. NDE staff must be certified to ASNT-TC-1A levels I, II or III depending on the responsibilities that they have on the project. The day-to-day inspection of production welds lies with the WI. Note that API 1104 does not identify the role of the RP 1169 PI – it was written long before that role was identified.

Section 9: NDE acceptance criteria

Section 9 is where API 1104 deviates from the established practice of most pressure equipment codes, leaning a little towards the more flexible approach of structural steelwork standards. Welding indications (using the AWS 3.0 definitions as we have seen) are divided into those that are detected by radiographic testing (RT), ultrasonic testing (UT), magnetic testing (MT), penetrant testing (PT) or visual testing (VT) techniques and acceptance levels are given for each. They are simple enough to understand and classify, helped by the straightforward arrangement of most pipeline weld joints – a uniform bevel butt joint with no complicated geometry. Figure 9.5 shows the abbreviations used.

Some of the weld indication acceptance limits vary with the diameter of pipe and most are related to the pipe wall thickness. As a general principle, if the wall thicknesses being joined are different then it is the *thinner* one that is used as the governing thickness when assessing the indication acceptance criteria. A quick look at the acceptance criteria of API 1104 Section 9 will show you that some 'robust' indications can still be accepted, as they are not considered a significant risk to integrity. This is the proven principle of API 1104.

FIG 9.5
Some API 1104 (section 9) indication definitions

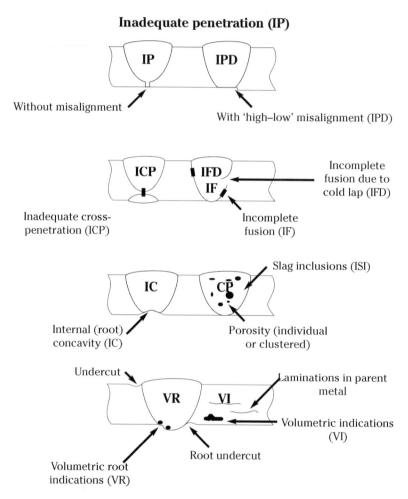

Ref: See API 1104 section 9 for indication acceptance levels found by RT, UT, MT, PT

Section 10: Repair of weld defects

You have to watch out for the hidden technical agenda here. Repairs to welds during fabrication are not unusual in any engineering component: even properly qualified welders make mistakes with machine settings, travel speeds and similar. With pipeline construction, however, the situation is complicated by the fact that the butt welding of one pipe spool to its neighbour is a repetitive and continuous process, frequently using semi-automatic welding techniques. While these two factors give the advantage of repeatability in the process (essentially a good thing for minimising defects) it also gives the opportunity for *systemic* problems to develop with the welding process if you are not careful.

The occurrence of systemic, repeated problems of the same type is not helped by the on-site activity of the welding activity itself. Poor site and weather conditions, problems with material supply to remote areas of the world and changes to personnel can all combine to cause problems with welding, heat treatment and NDE that can repeat themselves if not found quickly.

Pipeline contractors employ WIs to provide full-time monitoring on site, as we saw defined in RP 1169 annex D. The reality of minimising defects in site pipeline welding is that WIs have the capacity to be either part of the solution or, of course, part of the problem. It is not exactly unknown for WIs to have their technical objectivity overruled by construction schedule issues and cost decisions. Convenience and wishful thinking are common partners in some cultures of the world, and responsible for many of the seemingly obvious problems that have occurred on pipeline construction projects.

Now we see the role of the RP 1169 PI in weld inspection: it is to identify and prevent systemic welding problems occurring and repeating themselves (as they do) over the continuous 24/7 schedule of a pipeline programme. It's the systemic problems that are the important ones – the difficulty is to differentiate them from individual non-repeated occurrences that are properly identified and corrected by the WIs before they have the chance to grow into repeated occurrences.

API 1104 section 10 *Repair and removal of weld defects* gives a place to start. It specifies three types of repair that constitute a higher risk. These are

- repairs to weld cracks
- back-weld repairs (meaning repairs to be root side of a groove weld (see definition 3.1.2))

- double repairs (a repair to a repair because the first one wasn't done properly).

These make sense. Taken together they make-up perhaps 75% + of the reasons behind pipeline weld failure in service. The principle of API 1104 is that higher-risk welding repairs must be covered by qualified WPSs/PQRs and welder qualifications, much the same as for a new weld (see Figure 9.2). In addition, on a repair, there are various technical constraints on what is allowed to be done – the maximum weld length that can be excavated and re-welded is a typical example. It is an important responsibility of the PI, using whatever surveillance is required, to ensure that individual repairs that do occur are diagnosed and performed correctly, preventing them from growing into the repeatable systemic problem previously described.

Repairs – the technical objectives
Given the tendency of poor weld repairs to be a contributory factor to a lot of pressure and structural equipment failures (not just pipelines) it is worth looking at exactly what it is that causes the problems. Setting aside, conveniently, the problem of using the incorrect welding materials/consumables, the objectives are straightforward.

- To prevent excessive *hardness* of the repair.
- To guard against the repairs being *too brittle* (closely related to its hardness).
- To make sure the repair is not too soft (and so weak).
- To eliminate the defect that caused the need for the repair in the first place.

From an inspection viewpoint, the first two objectives – hardness and brittleness – are similar. High material hardness (Vickers hardness (HV) or Brinell hardness (HB)) in a steel is closely linked to it being brittle, rather than tough (the opposite). Assuming the material has the correct percentage carbon equivalent level (i.e. it is not too high), then the reasons for excessive hardness in a construction repair weld are

- too much heat input
- not enough post-weld heat treatment (PWHT)
- too-quick cooling after welding (or PWHT if it has it).

Add all three together and the situation gets worse. Under field welding conditions all of these can feasibly occur in butt welds. Fillet welds used for reinforcing sleeves and attachments can have the extra problem of difficulty with physical accessibility of the weld. Welders may turn up

FIG 9.6
Hardness testing of repair welds

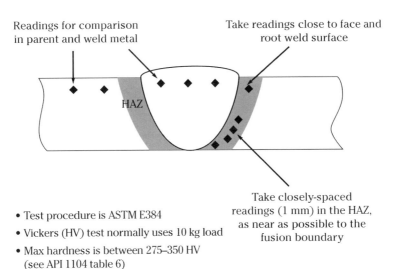

- Test procedure is ASTM E384
- Vickers (HV) test normally uses 10 kg load
- Max hardness is between 275–350 HV (see API 1104 table 6)

the welding current, increasing the heat input to get round the problem, at the expense of going outside the qualified variables of the repair WPS.

In a field welding situation, surface hardness tests are the main diagnostic tool available to a PI. Details are given in API 1104 (10.3.7.3) referencing ASTM E384, a well-established procedural document for hardness testing. Readings are taken across the weld metal and heat-affected zone (HAZ), with particular reference to the fusion zone between the two. Figure 9.6 shows the technical details.

Section 11: NDE procedures

Section 11 provides basic details on NDE techniques by RT, UT, PT and MT. It is well-established information, much the same as in ASME V, but in abridged form. Emphasis is placed on

- the need for an approved procedure for each of the techniques
- details of the procedures themselves (as per ASME V)
- qualification/certification requirements for the NDE technicians doing the tests.

For those PIs from an NDE background, the technical information in this section should prove fairly easy to understand. For API 1169 exam purposes, unlike the API SIFE exam, there is not huge emphasis placed on the need for a PI to have detailed NDE knowledge. It is sufficient to be able to carry out overall surveillance and verification that the contractor's NDE personnel are doing the job properly.

9.3 API 1104 sample questions

Once you have read through this chapter and looked up the reference points in API 1104 itself, you can try the sample question set that follows. If you don't have a basic knowledge of NDE then you should put additional effort into reading sections 8 and 11 of API 1104, looking up the meaning of any terms you do not understand. If you are weak on the mechanical properties of steels, then API RP 577 *Welding Inspection and Metallurgy* contains useful background reading. This document is not in the API 1169 BoK but does appear in the BoKs for SIFE and other API ICP examinations.

Once you have done the reading, try question set 9.1 (closed-book first) to see how you do, then look up in API 1104 to see where the correct answers came from.

Question set 9.1: API 1104 pipeline welding

Q1. API 1104: qualification of welders

The purpose of a welder qualification test is to determine the ability of welders to make sound butt or fillet welds using previously qualified procedures. A welder shall be qualified to the requirements of API 1104 if they are successfully qualified on test pieces comprising

(a) Segments of pipe nipples or full-size pipe nipples ☐
(b) Pipe or flat plate joints ☐
(c) Full-size pipe nipples only ☐
(d) Any joint configuration of representative base material and wall thickness ☐

Q2. API 1104: welding operations

During welding of API 1104 pipelines, destructive tests on production welds are

(a) Only required if specified by the pipeline owner/user/constructor ☐
(b) Performed on at least 1% of each welder's work output ☐
(c) Prohibited ☐
(d) Mandatory ☐

Q3. API 1104: NDE personnel

For an API 1104 pipeline welding programme, all levels of NDE personnel shall

(a) Have at least 3 years' experience ☐
(b) Be recertified at least every 5 years ☐
(c) Have at least 5 years' experience ☐
(d) Be recertified at least every 3 years ☐

Q4. API 1104: welding terminology

Which of the following statements best defines an arc welding process for pipelines that is semi-automatic?

(a) The operator controls the arc between a covered electrode and the weld pool ☐
(b) The advance of the welding is manually controlled and the filler metal feed is being controlled by the equipment ☐
(c) The pipe or assembly is not rotating while the weld is being deposited ☐
(d) The advance of the welding and the speed of the filler metal are controlled by the equipment ☐

Q5. API 1104: welding terminology

An API 1104 pipeline weld in which the pipe is rotated while the weld is being deposited near the top centre position is a

(a) Roll weld ☐
(b) Flat weld ☐
(c) Position weld ☐
(d) Submerged arc weld ☐

Q6. API 1104: welding procedures

When welding pipelines to API 1104, base material, welding processes and weld position are

(a) Essential variables ☐
(b) Not included in the WPS ☐
(c) Supplementary variables ☐
(d) Chosen by the inspector ☐

Q7. API 1104: welding materials

Before starting production welding of pipelines to API 1104, the WPS shall be qualified and the quality of the weld determined by

(a) Volumetric NDE ☐
(b) Destructive testing ☐
(c) Chemical analysis ☐
(d) Fatigue testing ☐

Q8. API 1104: welding definitions

Which of the following terms would be used to define a discontinuity or irregularity that is detectable by the methods outlined in API 1104?

(a) An indication ☐
(b) An imperfection ☐
(c) A defect ☐
(d) An indication and defect ☐

Q9. API 1104: weld acceptance criteria

When assessing pipeline welding to API 1104, transverse indications

(a) Are defined as three-dimensional indications ☐
(b) Are defects ☐
(c) Have their greatest dimension along the weld ☐
(d) Have their greatest dimension across the weld ☐

Q10. API 1104: weld defect acceptance criteria

When assessing a pipeline weld to API 1104, a subsurface imperfection between the first inside (root) weld pass and the first outside weld pass caused by inadequate penetration of the vertical land faces is referred to as

(a) Internal Cavity (IC) ☐
(b) Incomplete Fusion (IF) ☐
(c) Inadequate Cross Penetration (ICP) ☐
(d) Inadequate Penetration Due to High–Low (IPD) ☐

Chapter 10

Pipeline defects

10.1 Indications and defects

Anticipating the types of defects that occur during site assembly welding of pipeline sections takes up quite a lot of API 1104 *Welding of Pipelines and Related Facilities*. Ultimately, the task is not that difficult – there is little in the code that is technically new, so there are not many real surprises to be had. Arguably, all defects in pipeline welding techniques should have been eliminated years ago, as victims of improvements in materials, weld consumables and the reliability of the welding processes used. Technologically, this is the case – it's just that the practicalities of welding on a spread-out construction site with its variability of conditions, project pressures and personnel sometimes come along to spoil the party.

API 1104, like many other construction codes, is a *utilitarian* document. Its role is to provide a practical solution to the activity of site pipeline welding, rather than the best that is technologically available. It doesn't claim, therefore, to be a document of technical excellence.

Section 9 of API 1104 (*Acceptance standards for NDT*) demonstrates the practical nature of the code by listing all the relevant imperfections that occur during the welding process and giving levels of their acceptance or rejection. We saw this back in Figure 9.1 of this book where the terminology was set out as

- an **indication** – a finding made by NDT that may or may not be real
- an **imperfection** – an indication that is real (physically exists) but is within code (API 1104) acceptance levels and so is considered *acceptable*
- a **defect** – an indication (real) that is outside code acceptance levels and therefore is *rejectable*.

Why some indications can be considered acceptable

Behind this lies a surprising amount of technical argument. Let's see if I can explain. Looking at a weld from a metallurgist's perceptive, no weld or the parent material it joins can be actually considered *perfect*. All are stacked full of imperfections at the molecular or granular (macro) scale – it's only their size and shape that are worth arguing about. Metallurgists will name these as dislocations (edge or step-type), intergranular or trans-granular features or other more complicated terms. Once these get large enough to be visible and physically relevant in some way, then along come welding engineers who translate them into the names of physical imperfections such as lack of fusion, lack of penetration, inclusions and suchlike. All sounds good.

The only problem so far is that once these imperfections – large or small, named or un-named – are found, classified and recorded, no-one really yet knows exactly how they will affect the integrity of the engineering component in which they reside. In this case our overland pipeline. This is the job of the pipeline designer or, to be more precise, the 72 or so of them that inhabit the code committee charged with deciding these things. Seventy-two people, plus participating members of various code sub-committees will always have difficulty in reaching technical agreement on just about anything. Even simple technical arguments about whether, for example, weld undercut acts more as a type of pipe wall thinning than it does as an initiator of cracks during a hydrostatic test can be the subject of wide technical opinion, informed and otherwise. This doesn't mean that consensus on what is a safe size of indication is impossible, just that it is always elusive.

One solution to this lies in the use of hindsight. Pipelines have a long and fairly well-documented history. If it can be established that a pipeline operated successfully at a pressure of X psi without failure while containing an imperfection of length Y, then it's a fair starting assumption that others will do the same. All you need now is a bit of validation using burst tests of pipe sections containing known sizes of imperfections to increase the confidence level among any doubters. As a bonus, once the safe imperfection is adopted it's easy to develop a theory, or extend some existing one, to fit the empirical findings.

Not all code acceptance criteria are decided exactly like this, but there is always some close relationship between previous empirical results and the application of theories and analyses. Even in the current age of computer analysis this balance still applies, and pipelines fit the model

FIG 10.1
Code defect acceptance criteria
– Where do they come from? –

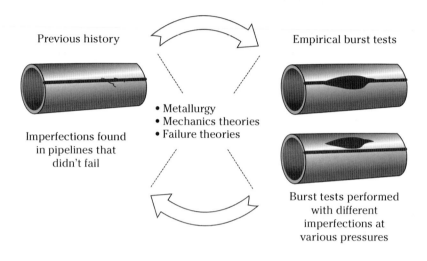

Previous history

Imperfections found in pipelines that didn't fail

- Metallurgy
- Mechanics theories
- Failure theories

Empirical burst tests

Burst tests performed with different imperfections at various pressures

For pipelines the situation is simplified:
- Pipelines have a uniform wall thickness
- Simple bevel butt welds
- C–Mn steel is one of the most predictable materials
- API 1104 classifies all pipelines ≥2.375 in. diameter as 'large diameter' with similar failure mechanics.

better than most other pressure equipment components. Figure 10.1 shows this in illustrated form.

10.2 The acceptance criteria of API 1104

Technically, these are straightforward, but presented in a particular way. Section 9 of API 1104 divides imperfections into categories based on the NDE technique by which they are found (i.e. RT, PT, MT, UT and VT). This is a slightly artificial method of classification as many of the imperfections can be found by more than one of the NDE methods. Practically, however, it doesn't make a lot of difference.

Figure 10.2 shows a summary of the way the code provides the information on acceptance criteria. Note the following points.

- The abbreviations used for some imperfections (IP, IC, IPD etc.) are different to those used in other pressure equipment codes.
- For many imperfection types, acceptance criteria are based on a combination of the dimension (length) of the longest imperfection and their aggregate lengths measured over 12 inches or a limit of 8% of the total weld length.
- Linear cracks are always rejectable.
- Not all imperfection types are directly mentioned by API 1104. Arc strikes, weld spatter, underfill and specific types of micro-cracking (hydrogen cracking, centreline solidification cracking etc.) are just lumped together under the term 'cracking'. This is evidence of the practical rather than academic approach of API 1104 – not necessarily a bad thing.

Note that there is an attempt in the sections on burn-through and slag inclusions to link the acceptance criteria to whether they are found in a small- or large-diameter pipe. This is a slightly strange one as large-diameter pipe (see 9.3.7.2) is defined as having a diameter ≥2.375 inches (60.3 mm) – an impractically small size for a pipeline. There is also little recognition in any of the imperfection types of pipeline wall thickness. API 5L pipeline material is available in schedule sizes up to 2 inches (50.8 mm) thick and the thicker a material is the greater the influence of weld imperfections on crack propagation. It is perfectly possible to have a weld imperfection that causes no problem in thin ($<\frac{1}{2}$ inch) material but can assist crack propagation in thicker section material. This is due to the greater heat input generated when welding thick sections, plus the fact that the thick parent material is inherently more brittle than thin sections of the same material to start with.

The influence of welding technique

Circumferential (or 'girth') welds for joining pipeline sections on site are a straightforward welding technique. They are simple single- or double-bevel butt welds on material of constant thickness with, on larger-diameter pipelines, unobstructed access to the inside and outside. The weld joint preparation is normally pre-bevelled in the pipe manufacturing works and is a simple bevel with a small root land and root gap. Welding these together on site is done in one of the following three ways.

- **Manual welding**. Sections are welded manually, from the outside. One welder on each side of the pipe welds their half of the circumferential

FIG 10.2
RP 1104 defect acceptance

Indication type and its method of detection	Criteria used to assess if it is a defect	RP 1104 reference
Found by RT (section 9.3)		
Inadequate (root) penetration (IP)		9.3.1
		9.3.2
Inadequate cross (inter-run) penetration (ICP)		9.3.3
Incomplete fusion (IF)		9.3.4
Incomplete fusion/cold lap (IFD)		9.3.5
Internal (root) concavity (IC)	No length restriction	9.3.6
Burn-through (BT)	• Pipe diameter/wall thickness	
	• Maximum dimension	
	• Aggregate length	9.3.7
Slag inclusions	• Pipe diameter/wall thickness	
	• Length and aggregate length	
	• Width of indication	9.3.8
Porosity (isolated)	• Size of largest bore	
	• Distribution of bore	9.3.9.2
Cluster porosity (CP)	• Diameter of cluster	
	• Aggregate length of cluster	9.3.9.3
Hollow bead porosity (HB)	• Length of indication	
	• Aggregate length of indication	
	• Spacing of indications	9.3.9.4
Cracks	All cracks in welds are considered defects	9.3.10
Undercut (EU, IU)	• Aggregate length of indications	
	• Location adjacent to cover pass (EU) or root pass (IU)	9.3.11
Found by MT (section 9.4)		
Linear indications	• Length dimension	9.4.2
Rounded indications	• Assessed as porosity	9.4.2
Found by PT (section 9.5)		
Linear indications	• Assessed as (IF) on length	9.5.2
Rounded indications	• Assessed as porosity	9.5.2
Found by UT (section 9.6)		
Linear indications (surface or buried)	• Indication length	
	• Aggregate length	9.6.1.2
Transverse indications	• Assessed as volumetric indications (VI)	9.6.1.3
Volumetric individual indications (VI)	• Maximum indication length	9.6.1.4
Volumetric cluster indications (VC)	• Maximum indication length	9.6.2.5
Volumetric root indications (VR)	• Maximum indication length	
	• Aggregate length	9.6.2.7

joint, starting at the top and progressing *downwards* (see Photo 10.1). This downward technique is the quickest and most efficient at avoiding defects, compared with the weaving technique needed for upwards welding.
- **Double-ending**. Here, a double-length pipeline section is created by submerged-arc welding (SAW) two sections together in a moveable site welding facility. SAW is a semi-automatic process that, once properly set up, can produce welds with almost zero incidence of defects. The pipe needs to be rotated when welding owing to the nature of the SAW technique. The double-length section is then taken to the edge of the trench where it is welded to the next double-length section, and so on.
- **Automatic orbital welding**. Done correctly, this is the quickest and most reliable process to use for long continuous pipeline projects. Adjacent pipe sections are held in the correct alignment by a special clamp. This incorporates a steel guide ring that fits round the full circumference of the outside of the pipe. An automatic welding machine is then clamped to the pipe using magnetic fixings. Once started, the machine moves orbitally around the pipe under its own power, guided by the guide ring. The pipe stays still so the varying angles, rotation and wire feed speeds etc. are all done automatically by the machine as it travels round. Sometimes, the weld root run is done manually and then the machine is used for the fill passes. In others, the root pass can also be done automatically.

Technically speaking, the list of weld imperfections and their acceptance criteria given in API 1104 can apply to any of these welding techniques. In practice, manual welding tends to produce the most defects simply due to the involvement of people in the welding process. Coded welders still make mistakes and take shortcuts. Automatic welding processes depend heavily on the attention paid to their set-up. If set up correctly they can produce imperfection-free welds time after time. Alternatively, if they are set up wrongly they will reproduce defects in the same way, sometimes without them being immediately obvious. This is not a common occurrence but has happened; weld centreline cracking and lack of root penetration have been built into long pipeline sections, requiring major repair or replacement.

Solving welding problems

The good news is that this is not job of the pipeline inspector (PI). The pipeline contractor has a site welding inspector (recognised by RP 1169

Photo 10.1 Manual pipeline welding (downhill) (photo courtesy Bigstock)

remember) to find defects, and welding engineers to decide and qualify the techniques to be used. Between them, they have the main responsibility to solve welding problems, leaving the PI to provide high-level monitoring of API 1104 compliance, along with all the other responsibilities set out in RP 1169. The exam BoK reinforces this view, with weld inspection playing a vital, but still small, part.

10.3 Pipeline surface distortions

Checking for distortion or damage to pipeline sections is one of the physical checking roles of the PI. These can occur either during manufacture, loading/unloading during transport to the site or storage area or during hauling and lowering-in at the RoW itself. Specific distortions can be also introduced during the welding process. The problem with distortions (of any type) is their ability to produce a stress concentration in the pipe wall when it is under pressure, either during hydrostatic testing or later in service. Pipeline construction codes and federal regulations therefore contain requirements to stop this.

Distortion terminology

The construction codes ASME B31.4 and ASME B31.8 classify pipeline distortion under the loose term of '*surface requirements*'. Within this they are divided into the following.

- **Dents and bulges**. These are distortions to the circular profile of the pipe but without significant reduction in wall thickness. They may be of smooth profile or sharp-edged.
- **Gouges and grooves**. These are features that reduce the wall thickness in some way. They are normally caused by the mechanical action of cutting tools or grinding wheels.
- **Notches**. Mechanical (grinding) notches are similar to gouges and grooves but are often smaller, sharper and more likely to initiate cracks. ASME B31.8 also uses the term *metallurgical notch* (821.2.4 (d)) to explain stress concentrations resulting from local and excessive heating of the metal surface, for example from welding arc burns.

Different terms can also be used for different types of profile distortion. Figure 10.3 shows these for buckles, knobs, out-of-roundness, peaking, rucks, wrinkles and misalignment.

The need for acceptance limits

Designers, being designers, use equations for pipeline design that assume that the wall of the pipe is a perfectly circular thin 'shell'. This perfect circle assumption allows the stress in the shell to be classified as a *membrane* stress. Think of it as a nice uniform stress as the circle is restrained from expanding evenly outwards under the influence of the pressure acting evenly over all the internal surface. The simple hoop stress equation allows easy calculation of the size of this stress (S) and, by re-arranging the formula, what thickness of material is needed to resist the internal pressure. The hoop stress is given by

$$S = \frac{PD}{2t}$$

in which P is the internal pressure, D the pipe dimeter and t the pipe wall thickness. Rearranging this leads to a required wall thickness of

$$t = \frac{PD}{2S}$$

where S is the stress in the pipe *allowed by* the code.

FIG 10.3
Pipeline surface distortions

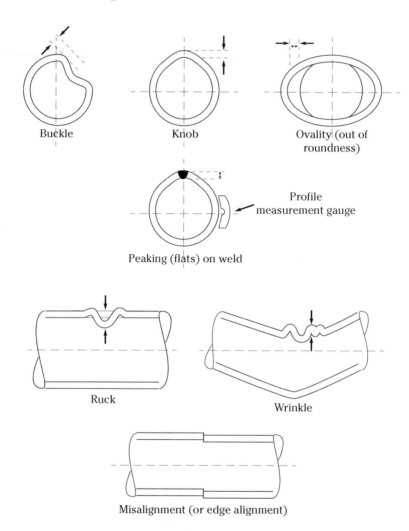

Unfortunately, this nice neat world of membrane stress only applies if the pipe is infinitely thin (which it is not) and perfectly circular. The finite thickness of a real pipe can be dealt with by adding a few second-order factors into the equation. Any lack of circularity, however, soon makes the hoop stress equation completely inaccurate. This is because the deviation of the form from a perfect circle introduces bending stress into the material. This superimposes itself on the (now distorted) hoop stress field, making the real stress next to impossible to calculate accurately. This condition is known as being *statically indeterminate*.

To reduce the chance of failures, design codes introduce acceptance limits on the amount of profile distortion allowed. Safe levels are determined from a combination of historical experience, empirical burst tests and advanced stress analysis, as we saw in Figure 10.1.

What are the acceptance limits for new gas pipelines?

Figure 10.4 shows the acceptance limits specified by the gas pipeline code ASME B31.8 (NPS means nominal pipe size). They apply to pipelines operating at a hoop stress $>40\%$ SMYS. Note how any distortion that is sharp-edged is automatically rejectable. These distortions can result from poor-quality manufacture, mechanical damage or a combination of both.

The role of the PI

The best time for a PI to start to look for pipeline distortion is when the pipe sections arrive in the unloading/storage area. Most can be spotted visually but, for some, a special profile gauge is required. These are made to a specific pipeline diameter and laser-cut out of thin sheet steel. Distortion problems with pipeline sections tend to be linked to a single batch or heat run at the manufacturing works so, once one is found, others from the same batch should be checked for the same problem. Ovality and peaking are the most common distortions produced during manufacture.

The next stage of inspection is to find any distortion caused by hauling and stringing the pipe section along the edge of the trench. On sites with difficult terrain, the sections may have been dragged over rocks – the main cause of site dents. Cold bending (by a mobile bending machine) is required to ensure the pipe sits in any slightly curved trenches. It is a regulation requirement that the pipe must be laid clear of the sides of the trench, so the need for cold site-bending is not

FIG 10.4
Pipeline dents: acceptance limits

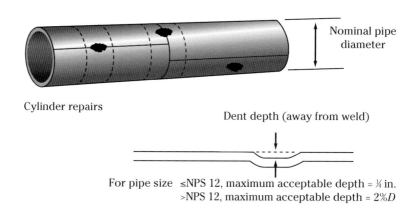

All dents on welds must be repaired by cutting out a cylinder and rewelding

Nominal pipe diameter

Cylinder repairs

Dent depth (away from weld)

For pipe size ≤NPS 12, maximum acceptable depth = ¼ in.
>NPS 12, maximum acceptable depth = 2%D

Notes
- These show ASME B31.8 gas pipelines limits (same as 49 CFR 192)
- All dents with sharp edges are rejectable and must be repaired
- Repairs can only be made by cutting and replacing a cylinder – not hammering out
- Limits shown are for hoop stress >40% SMYS

unusual. Excessive bending can result in the ruck or wrinkle defects shown in Figure 10.3.

Assuming the pipeline sections are of consistent diameter and wall thickness (within the dimensional limits of API 5L) then the welding process should not cause any new 'circularity' distortions to occur. Problems are most likely to be pipe wall misalignment caused by loose or inaccurately placed alignment clamps. Setting these up is a manual process so is always open to errors occurring. Out-of-specification differences in diameter or wall thickness make this problem more likely. These errors can be spotted visually during or after welding is completed – hence the need for PI surveillance before coating and lowering-in. RP 1169 and the exam BoK cover this surveillance as an essential role of the PI.

Chapter 11

Pipeline pressure testing: API RP 1110

11.1 Pressure testing – what's it all about?

Pressure testing of a pipeline to check its integrity before it goes into service is one of the most important stages of the construction programme. It is also one of the most important roles of the RP 1169 pipeline inspector (PI), combining, as it does, responsibilities for surveillance, verification and reporting in both technical compliance and safety considerations. We saw in Chapter 5 how this combined safety and technical role is RP 1169's view of the role of the PI.

Does pressure testing make good API 1169 exam questions?

It certainly does. Both the technical and safety aspects of pressure testing carry some 'serious' requirements that form real issues in any pipeline construction project, making it an important area for verification that everything is done correctly. This makes it a key component of the exam body of knowledge (BoK) and a valid source of exam questions.

Which codes cover pressure testing?

The main one referenced in the API 1169 Code Effectivity List (see Figure 5.1) is API RP 1110 *Pressure Testing of Steel Pipelines*. This is in the closed-book question part of the exam. In reality, the situation is not that simple because requirements relating to pressure testing are spread out over five or six of the BoK documents, covering both technical and safety aspects of the testing procedure. Figure 11.1 shows the details – note how the PI's responsibilities are drawn from four areas of the code list.

- **API RP 1169** defines the *PI's responsibilities* in verifying and reporting on pressure tests.
- **ASME B31.4 and ASME B31.8** confirm *construction code requirements* and procedural details of the tests themselves.
- **API RP 1104** gives *technical objectives* and more procedural details of the tests themselves.
- **INGAA CS-S-9** specifies *safety requirements* for pressure tests supported by federal regulations 49 CFR 192 and 49 CFR 195, putting them in the general context of construction site safety of personnel.

Of the documents shown in Figure 11.1, only the CFR documents are in the open-book questions part of the exam. The others are closed-book. Some information, mainly safety principles and requirements are duplicated in several documents, but this is no real problem. The sample questions at the end of this chapter contain examples from across the spread of these documents.

The special nature of pipeline pressure tests

The nature of construction and the long physical length of overland pipelines make their pressure testing different from those performed on general static pressure equipment such as vessels, pipework and heat exchangers. In most cases it makes the technical and procedural aspects *more difficult* rather than easier. Costs are also higher, and safety precautions more arduous and difficult to enforce under site conditions.

The technical complexity of pipeline pressure testing comes from the necessity to decide and use the correct test pressure. Unlike vessels, where the construction code specifies a single minimum hydrostatic or pneumatic test pressure, the test pressure of a pipeline has to be derived using consideration of its design limits, how much stress it will experience in use and for how long, and its location class (i.e. whether its route passes near people, populated areas, transport facilities etc.). Both the minimum and maximum allowable test pressures are affected by these.

Procedural difficulties in performing the test are caused by the long physical length of a pipeline and its remote location. For anything longer than a few kilometres, the pipeline is normally pressure-tested in sections, requiring the provision of temporary isolations.

Other difficulties come from

- the need to procure and dispose of large quantities of treated water

Pipeline pressure testing: API RP 1110

FIG 11.1
Pipeline pressure testing
– The documents involved –

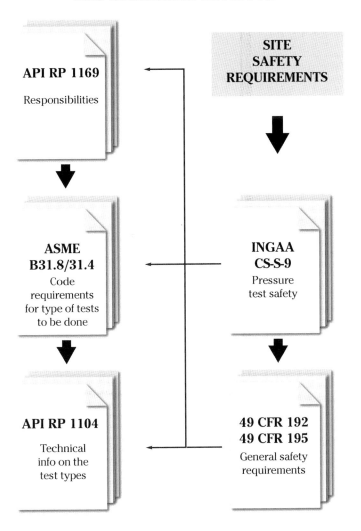

- the requirement for environmental permits
- the difficulty of bleeding all the air out of a pipeline that is being tested using water; this causes a safety risk
- general site safety problems with excluding personnel from dangerous test areas across a large construction site containing hundreds of people.

These 'difficulty points' covering both technical and safety areas fall squarely within the remit of the PI. All the risks involved are covered by the safety-oriented code documents shown in Figure 11.1. The PI is not responsible for performing any of the test activities but has the RP-1169-specified role to verify they are done correctly and to document the results to the requirements of the pipeline owner/operator.

11.2 The pressure test itself: RP 1110 content

RP 1110 is a short document (25 pages or so) divided up chronologically into the various steps necessary to propose and perform the pressure test. Figure 11.2 shows the basic steps involved; these are reasonably common to the other documents in the pressure testing 'set'.

What is RP 1110's guiding principle?

To bring some order to the principles and practice of pipeline pressure testing, RP 1110 states its guiding technical principle in its introduction and section 4. This is consistent with its role as a RP (recommended practice) document in providing technical justification and explanation to the mandatory statements of construction codes (ASME B31.4 and ASME B31.8) and all the safety-related documents that are only concerned with safety.

The guiding principle of RP 1110 is

> *To detect and eliminate stable and time-dependent anomalies, and so verify the integrity of the pipeline*

Most of this is hidden away in section 4.1.2 of the document, embedded within a bunch of other statements about objectives being met, things being done properly and everyone going home safely. The real technical meat of it, however, is clear to see, and gives purpose and structure to the rest of the RP 1110 document that surrounds it.

FIG 11.2
The basic steps of a pipeline pressure testing procedure

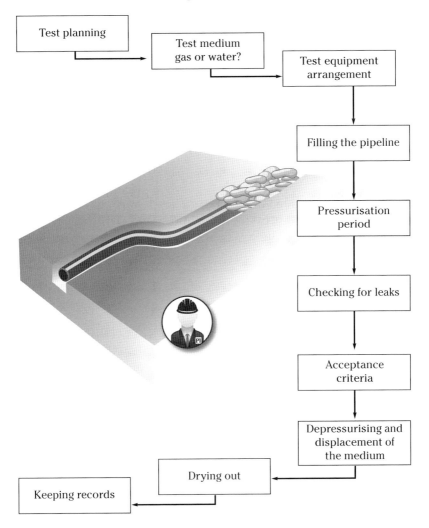

Did you spot the technical keywords?

There were three: '*anomalies*' and two terms used to describe them '*stable*' and '*time-dependent*'. These are used specifically in relation to pipeline testing, rather than other pressure equipment. Their importance lies in the way that the method of detecting and eliminating them (as the RP 1110 guiding principle) requires different types of pressure test. These are termed strength test, spike test and leak test – have a quick look now at Figure 11.3 to see the difference.

The two types of anomalies

Anomalies is the pipeline world's name for imperfections in the structure of a parent material or weld that have the ability to affect its integrity. In API code-speak they would be called *defects* if they were large enough to fall outside any acceptance criteria given in the construction code. RP 1110 divides them into two types, stable and time-dependent, defined as follows.

- *Stable anomalies* are currently too small to cause a failure risk and are expected to stay that way over time.
- *Time-dependent anomalies*, while not necessarily affecting integrity at the moment, are expected to grow over time. This is called sub-critical growth and will continue, if allowed, to a point where they are large enough (*critical size*) to cause a failure risk.

Stable anomalies cause the least problem because if they can survive a pressure test at an elevated test pressure without a resulting failure then it has been proved they are non-critical and won't cause failure at the lower operating pressure that will be seen in service. The accepted way of detecting stable anomalies is using the *strength test* type of pressure test.

Time-dependent anomalies present the problem that, although they might pass a strength test, they are expected to grow over time. This means that the strength test would have to be repeated at intervals in the future to check whether or not they had grown to a critical size and were at a risk of causing failure. Real-world pipeline flaws such as material imperfections, fatigue-crack initiators and stress corrosion-cracking behave in this way.

The best way of detecting/eliminating these types of anomalies is to do the pressure test at a *higher pressure* than that for the strength test. This is the *spike test*. This more strenuous test detects those anomalies that are big enough (now) to grow to critical size over a projected

FIG 11.3
The three types of RP1110 pipeline pressure test

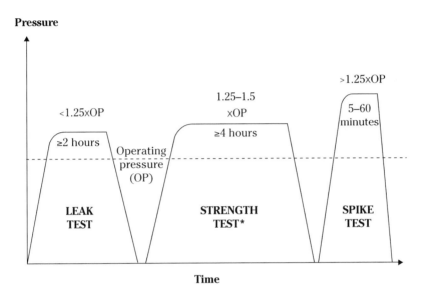

*See ASME B31.8 for code time/pressure requirements

Acceptance criteria (as per RP 1110)

Leak test: Any pressure variation can be explained

Strength test: No pipe failures or leaks occur

Spike test: No pipe failures occur

timescale, providing sufficient confidence to extend the period before it is necessary to do another strength test (or another spike test if the anomalies are still considered time-dependent).

The importance of pressure ratio

The term *pressure ratio* is simply the RP 1110 term for the ratio between the pressure in the pipeline section when it is pressure-tested and at its 'design' operating pressure.

This ratio is important to RP 1110 in defining what pipeline test and operating pressures actually are. It is used in defining the strength test, spike test and the other one in the group – the leak test. Now let's look at what these tests actually are.

$$\text{Pressure ratio} = \frac{\text{Test pressure}}{\text{Operating pressure}}$$

11.3 Pressure test types

The strength test

The purpose of the strength test, by detecting/eliminating stable anomalies, is to establish the *safe operating pressure* of the pipeline. Or, you can think of this in another way: it demonstrates the safety of the pipeline at the operating pressure that has been *planned for it*. They both mean the same.

This strength test is the one specified by the mandatory requirements of a pipeline construction code, such as ASME B31.8, or its equivalent for pressure vessels and pipework. ASME B31.8 specifies different test requirements for three gas pipeline design cases, based around how much stress the pipeline will see in service and the pipeline route location. The three cases are

- hoop stress ≥30% SMYS (specified minimum yield strength)
- hoop stress <30% SMYS and operating pressure ≥100 psi
- operating pressure <100 psi.

For the first of these design cases, i.e. hoop stress ≥30% SMYS, the pressure ratio for the strength test varies between 1.25 and 1.50. The differences are due to the location class (1 to 4), which require different safety factors to be incorporated to reflect the consequence of a failure occurring in service. The ASME B31.8 test duration is a minimum of 2 hours. ASME B31.4, which covers liquid pipelines, has more relaxed requirements due to the lower risks compared with gas.

Figure 11.3 shows the strength test with its minimum pressurisation time of 4 hours. The extended period allows sufficient time for pressure stabilisation, often a problem due to ambient temperature changes along the pipeline route. As this is a strength test then the test is acceptable under RP 1110 if it *does not fail* while tested to the correct pressure for sufficient time.

The spike test

A spike test (see Figure 11.3) is the best test for finding *time-dependent* anomalies. It uses a higher pressure ratio (>1.25) than a conventional strength test to detect a smaller size of sub-critical anomalies. It is of short duration (less than 1 hour) so that any sub-critical anomalies do not have enough time to grow under the increased stress. The actual pressure ratio used varies between pipeline operators but is typically around 1.4. Construction codes ASME B31.4 and ASME B31.8 do not specify a mandatory spike test, but do not prohibit it either, as long as specified maximum test pressures (see ASME B31.8 table 841.3.2-1) are not exceeded. Similar to a strength test, the acceptance criterion for the spike test is only that the pipeline should not fail during the test. Small leaks from flange joints etc., which can be explained, are not a cause for rejection.

Pressure reversal
Pressure reversal is an undesirable but misleadingly named occurrence that must be avoided during a spike test. It occurs when the high spike stress is maintained for too long, giving sub-critical anomalies time to grow larger under the stress. Even if they do not have time to grow to critical size and cause failure, they grow to nearer to the critical size than they were before the test, hence any future test must be done at a lower test pressure to stop them continuing to grow. What has essentially happened is that the test has *reduced* rather than increased the continuing pressure capability of the pipeline. This is the origin of the term *pressure reversal.*

The negative effects of pressure reversal can be minimised by ensuring that maximum agreed test pressures are not exceeded for any significant length of time. One role of the PI in this is to ensure that pressure-relief devices are correctly incorporated into the pumps and test loops to prevent overpressure occurring.

The leak test

As its name implies, this is just a test for through-wall leaks, without attempting to test for strength or sub-critical anomaly size in any way. Owing to the possible complex arrangement of pipe sections, valves, temporary blanks and so on, it is not actually necessary for the whole assembly to be leak-free for the test to be considered acceptable. Leaks that are detected and explained fully are not a justification for test failure.

Detecting leaks can be difficult over long pipeline sections. In most cases the pipeline trench will already have been backfilled so visible detection will not be possible. The most common method is to watch for unexplained pressure drops over the test period. The minimum test period suggested by RP 1110 is 2 hours but this is commonly extended to improve the chances of detecting pressure drops. Spurious pressure drops can be caused by

- ambient temperature variations
- dissolution of gas and water test media
- lifting of relief valves due to pressure transients.

These are only a cause for rejection of the test if they cannot be explained away. If gas is used for leak testing then sniffers can be used. This is an accurate method but can be unreliable due to weather conditions and environmental factors.

Pressure testing – gas or water?

Either can be used; each has its own advantages and disadvantages. Large quantities of water require extraction and disposal permits in some environments, and disposal is particularly difficult if it has been dosed with a corrosion inhibitor or wetting agent. For gas pipelines, the test water has to be completely removed from the pipeline (dewatering) before commissioning – this is done by multiple runs of a cleaning pig. The main advantage of pressure testing using water is safety in the event of a failure under pressure. The stored energy released by the water, which is virtually incompressible, is relatively low. In practice, trapped air in the undulating route of a long pipeline section can negate a lot of this benefit. This means that even hydrostatic (liquid) pressure tests on pipelines should not be treated as safe. Figure 11.4 shows the summary of safety issues common to RP 1169, RP 1104 and INGAA CS-S-9.

Pneumatic testing using air or some inert gas is the most dangerous test of them all. Gas is 200 times as compressible as water, causing huge

FIG 11.4
Pipeline pressure testing safety requirements

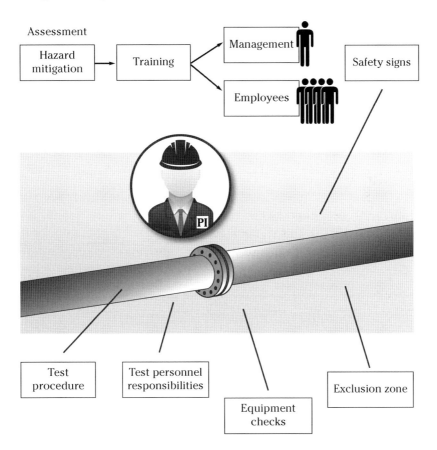

For full details see BoK reference documents:

- INGAA CS-S-9 *Pressure testing safety guidelines*
- 49 CFR 192 and 195
- CEPA/INGAA *Guide for pipeline construction inspectors*

release of stored energy if a component fails under test pressure. This results in catastrophic damage by flying missiles that can travel large distances: 400–800 m or more is not impossible. A large shock-wave follows, which can knock personnel into trenches, off scaffolding, or cause eye and ear damage.

For these reasons, pneumatic testing is treated very cautiously by all the safety-related codes that influence pressure testing. Hazard analysis, risk mitigation measures, exclusion zones, equipment pre-testing and all the safety-related steps in Figure 11.4 become much more important. Sounds like a good API 1169 exam question.

11.4 API 1110 sample questions

Given the technical importance of pressure testing in ensuring the integrity of a pipeline and the significant safety aspects involved, it makes a good topic for exam questions. Given the wide role of the PI and the multiple documents in the BoK covering safety, safety-related questions are almost guaranteed. Technical questions are sourced from the closed-book part of the BoK so you can expect them to be fairly straightforward. ASME B31.4 and ASME B31.8 are no longer a formal part of the BoK but are referred to as documents to be used 'for guidance only'. The required pressure testing requirements they contain, however, (particularly B31.8) are still important.

Have a try at question set 11.1.

Question set 11.1: RP 1110 pipeline pressure testing

Q1. RP 1169: PI duties during pipeline pressure testing

Which of the following defines the role of the pipeline inspector regarding the witnessing, signing off/on and recording the test plan for a pipeline pressure test?

(a) The local jurisdiction/statutory authority ☐
(b) The owner/operator ☐
(c) RP 1110 ☐
(d) RP 1169 ☐

Q2. RP 1110: pressure testing definition

A pressure test to RP 1110 designed to establish the operating pressure limit of a pipeline as required by code or regulation is an

(a) Operating pressure test ☐
(b) Proof pressure test ☐
(c) Spike pressure test ☐
(d) Strength pressure test ☐

Q3. RP 1110: pressure testing media safety

When may a pressure test using air or inert gas be performed in accordance with API RP 1110 to establish the operating pressure limit of a pipeline as required by code or regulation?

(a) When the pipe ROW does not cross rural areas ☐
(b) Only if the test is a short duration spike test ☐
(c) Only when a paper risk assessment says it's ok ☐
(d) Never; it is too dangerous ☐

Q4. RP 1110: pressure testing definition

A RP 1110 pressure test performed at a pressure ratio of less than 1.25 for a minimum duration of 2 hours is an

(a) Operating test ☐
(b) Spike test ☐
(c) Leak test ☐
(d) Strength test ☐

Q5. RP 1110: pipeline pressure testing

During pressure testing of a pipeline, the pressurisation to the agreed test pressure should be

(a) As slow as required to minimise fatigue stresses ☐
(b) Raised in 10% stages over a minimum of 2 hours ☐
(c) As fast as practical so as not to give anomalies time to grow ☐
(d) As slow as possible to equalise strains ☐

Q6. RP 1110: pressure testing procedure

When performing a pipeline pressure test to RP 1110 any excavated or uncovered segments of the pipe under test pressure should be

(a) Subjected to random sample NDE of circumferential welds ☐
(b) Backfilled prior to pressurisation ☐
(c) Left exposed to help identify leaks ☐
(d) Subjected to random sample NDE of longitudinal welds ☐

Q7. RP 1110: pipeline pressure testing: leaks

If a small leak is discovered during RP 1110 pressure testing of a pipeline, then

(a) The pressure should be reduced while locating the leak, then the test discontinued and repairs carried out ☐
(b) The leak should be located and the test continued to completion ☐
(c) The test should be discontinued immediately then repairs carried out ☐
(d) It should be recorded but the test can continue to completion ☐

Q8. RP 1110: pressure test effects

If a pipeline being hydraulically tested to RP 1110 exhibits continued leaks or failures that require the test to be repeated multiple times then

(a) The test pressure should be reduced ☐
(b) The pipeline should be scrapped ☐
(c) The test pressure should be increased to create a more stringent test ☐
(d) The test may be changed to a pneumatic test ☐

Q9. RP 1110: pressure testing procedure

When pressure testing a pipeline to API 1110, pressure recorder readings during the test should be compared with

(a) A deadweight tester ☐
(b) An additional set of calibrated gauges from an accredited test laboratory ☐
(c) P–V plots carried out on previous tests of the pipe spools in the manufacturing shop ☐
(d) An adiabatic comparator tester ☐

Q10. RP 1110: pressure test objective

When pressure testing a pipeline to RP 1110, detecting and eliminating time-dependent anomalies is best accomplished by using a

(a) High pressure test ratio and longer than normal test duration ☐
(b) Low pressure test ratio and shorter than normal test duration ☐
(c) High pressure test ratio ☐
(d) Low pressure test ratio ☐

Chapter 12

Federal pipeline regulations 49 CFR 192 and 49 CFR 195

Background: the legal requirements

Owing to the dangerous nature of pipelines carrying hazardous process fluids, particularly gas, onshore pipelines in the USA come under the far-reaching powers of the Pipeline and Hazardous Materials Safety Administration (PHMSA), a branch of the US Department of Transportation (DoT). Back in 1968, the National Gas Pipeline Safety Act was passed, requiring all pipeline operators to comply with a set of federal regulations providing minimum safety standards in all states of the USA. During this initiative, these subdivided into the series of 49 CFR (Code of Federal Regulations) documents. Two of these regulations are in the API 1169 exam BoK

- 49 CFR 192 covering *gas* pipelines and
- 49 CFR 195 covering *hazardous liquid* pipelines.

As a legal requirement in the USA, these regulations are enforced by the PHMSA under the provision of 49 CFR 190 *Pipeline Safety Enforcement*. As usual for official regulations they are written using the language *shall/will/must* for mandatory requirements and *should/could/may* for those that constitute non-mandatory guidance or good practice.

Application

Both sets of regulations apply to both onshore and offshore pipelines. They thus go outside the scope of RP 1169, which covers onshore pipelines only. They are targeted at the *owner/operator* of an installed pipeline, independent of who manufactured it. This differentiates them from the construction codes ASME B31.4 and ASME B31.8, which cover manufacturing only, without any reference to how they will be

FIG 12.1
Pipeline standards: The relationship between the ASME codes and CFR regulations

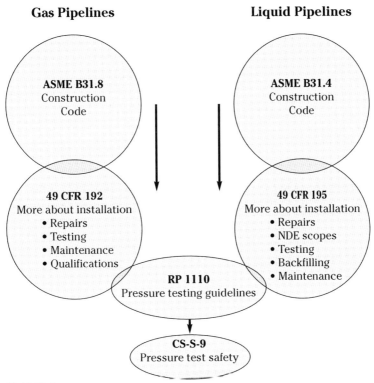

Note that:

- There are some areas of overlap between the documents. In the USA, CFR requirements take priority

- These documents cover both onshore and offshore pipelines
  The RP 1169 scope only covers *onshore* pipelines

installed as an operating pipeline. Figure 12.1 shows this relationship in a bit more detail. You can see the clear split between gas and liquid pipelines, but also the common requirements of the two documents regarding pressure testing (API RP 1110 and INGAA CS-S-9), both of which are in the API 1169 exam BoK.

The contents of 49 CFR 192 and 49 CFR 195

Only a few of the subparts of 49 CFR 192 and 49 CFR 195 are contained in the exam BoK. The full regulations are long and expansive (as regulations invariably are); 49 CFR 192 alone has a whopping 16 subparts. The only subparts we need to be interested in are

49 CFR 192 Gas pipelines	49 CFR 195 Hazardous liquid pipelines
Article 7: General	Article 2: Definitions
Subpart E: Welding	Article 3: Documents list
Subpart G: General construction	Subpart D: Construction (contains welding information)
Article 614: Damage prevention	Article 310: Pressure testing
Article 707: Line markers	Article 410: Line markers

Note that this content has been reduced from that in the pre-2017 API 1169 Code Effectivity List. Earlier ones contained information on personnel qualifications and pressure testing. These have now been removed as they are covered adequately by other documents in the BoK.

From a purely technical viewpoint the most useful information is that contained in subpart E *Welding* and subpart G *General construction* of 49 CFR 192. There is greater detail in these two subparts than in 49 CFR 195 for liquid pipelines where it is all squeezed into a single subpart D *Construction*.

Gas versus liquid pipeline requirements

Both 49 CFR 192 and 49 CFR 195 contain requirements that reflect the relative hazard levels of gas and liquid pipelines. 49 CFR 192, for gas pipelines, classifies them into risk types based on the level of hoop stress the pipeline experiences at its maximum operating pressure (MOP). This falls into three bands:

- hoop stress <20% SMYS (specified minimum yield strength of the material)

- hoop stress <30% SMYS
- hoop stress <40% SMYS.

Installations operating at the lower level are considered lower risk whereas, above this, the risk increases to a level where the component needs to be treated as a full 'coded item' (like a pressure vessel). It is then subject to similar code requirements in weld/welder qualification, defect acceptance levels and NDE/pressure testing. At <20% SMYS more relaxed rules can apply. Liquid pipelines under 49 CFR 195 do not have such a large spread of risk categories as the overall risk of liquids is considered lower.

If you compare both documents you will see that the construction requirements of 49 CFR 192 and ASME B31.8 are much the same. Cold bending is referenced in both and they have the same restrictions on wrinkle bends only being allowed for pipes operating at less than 30% SMYS. A wrinkle bend is an old site technique in which a pipe is cold-bent over a former, resulting in the inside (intrados) radius of the bend having a wrinkled or corrugated appearance. Wrinkles must be at least one pipe diameter long and the completed bend needs to be fairly uniform without any kinks. Figure 12.2 shows some repair requirements.

Installing the completed pipeline into its trench is covered by 49 CFR 192, but not by the construction code ASME B31.8. This makes sense as the scope of all ASME construction codes ends when the item leaves the manufacturer's works with all necessary testing and documentation completed. Requirements are given for

- **underground clearance**: how close the pipe can run to any unrelated underground structure (12 inches)
- **minimum cover**: the minimum covering of backfill required for different location classes based on the type of area and proximity to people and buildings.

Figure 12.3 shows some details.

49 CFR 195 for liquid pipelines

This is very much a slimmed down version of 49 CFR 192. It contains similar requirements for welding, but diversifies away from pipelines only and into general requirements for above-ground liquid storage tanks, pumps and valves. Strangely, it does not allow wrinkle bends to be used for hazardous liquid pipelines (whereas they can be used for gas). The first eight pages of the document are taken up with lists of

FIG 12.2
Repairs of gas pipeline spools
49 CFR 192 (ref 192.309)

For hoop stress >20% SMYS these dents must be removed

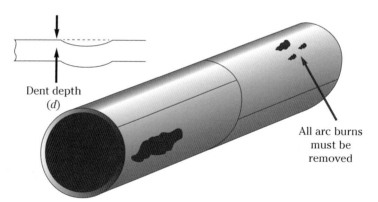

Dents containing stress concentrations (gouges, grooves, arcs, burns)

Dents affecting longditudinal or circumferential welds

For hoop stress >40% SMYS these must be removed

Dent depth (*d*)

All arc burns must be removed

Depth (*d*) of dents >¼ in. (6.25 mm) for pipe OD ≤12¾ in. (324 mm)
>2% OD for pipe OD >12¾ in.

FIG 12.3
Minimum clearance and coverage of gas pipelines: 49 CFR 192

Minimum Clearance

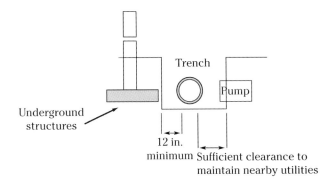

Minimum Coverage (Ref 192.327)

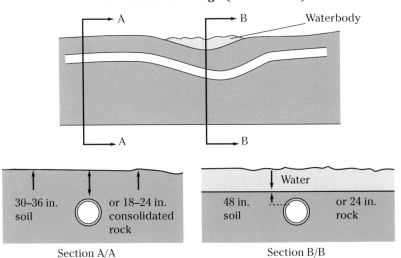

documents, terms and definitions, most of which are fairly general and coincide with the names and meanings given to them in the ASME B31 construction codes. Article 310 covers pressure test records – always a good topic for exam questions.

Finally, line marking

Both 49 CFR 192 and 49 CFR 195 are united in their view that permanent line markers must be placed along the routes of onshore pipelines so everyone knows that they are there. It is the consensus of the pipeline industry that the greatest single risk to an installed pipeline is from unauthorised excavation activities that puncture the pipeline. This can lead to serious environmental pollution and, in the worst cases, catastrophic explosions. Marker location, sizes, colour and wording are well described in the CFR, CGA and INGAA/CEPA documents.

Chapter 13

CGA best practice document 13.0 and INGAA crossing guidelines

13.1 CGA best practice document 13.0

This large 100+ page document is formally titled *The Definitive Guide for Underground Safety and Damage Prevention (13.0)*. It's therefore exclusively about excavation activities and ways to ensure that they don't cause damage to existing underground utilities. At its root is a system of *information, communication and co-operation* between all the involved parties ('stakeholders'). The idea is that a well-organised system will be much better than one that is fragmented, with uncertain rules about communication. There is no doubt that there have been frequent cases of this in the past in the pipeline industry, leading to some quite serious incidents.

What is the Common Ground Alliance (CGA)?

The CGA was formed as a result of work commissioned by the US Department of Transportation (DoT) and the Pipeline and Hazardous Materials Safety Administration (PHMSA). A group of 160 people from all stakeholder groups participated and put together the best practice document 13.0. It was – and still is – a document of consensus, with each of its points being endorsed by all the participating stakeholders. It contains lots of statements of generality but it also has clear detailed points as well.

What's the content of best practice document 13.0?

Looking at the chapter breakdown of the document gives you a good indication of how its contents are grouped. Figure 13.1 shows the breakdown. The first group of chapters (2 to 6) coincide with steps in

FIG 13.1
Common Ground Alliance (CGA)
– The breakdown of best practice document 13.0 –

Its title is **The Definitive Guide for Underground Utility Safety and Damage Prevention**

Chapter 2: Planning and design
Chapter 3: The one-call centre
Chapter 4: Locality and marking
Chapter 5: Excavation
Chapter 6: Mapping

These contain the information, communication and co-operation activities of the stakeholders.
See Figure 13.2 for the list of who they are.

Chapter 7: Compliance
Chapter 8: Public eductation and awareness
Chapter 9: Reporting and evaluation

More general 'community-type' requirements. Don't get too involved with them.

REMEMBER, IT'S ALL ABOUT 'CALL 811 BEFORE YOU DIG'

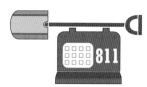

the pipeline construction project. They break down the activities required of the various stakeholders in a lot of detail. Chapters 7 to 9 cover more general aspects of linking the activity of pipeline excavation with local communities and wider society at large. Pipeline owner/operators take this sort of thing quite seriously; safety and environmental concerns are difficult to avoid in pipeline projects given the long distances and varied terrain they pass through.

The essential content of each of the document chapters (particularly 2 to 6) is set out in the following way.

- A short **practice statement** giving the principles of *what to do* for communication and co-operation to work properly. This is, at most, a sentence or two.

Each of these is followed by

- **practice descriptions** – lists of ways of *how* to do it.

Most of the 'practice description' lists are long and contain lots of information of little interest to the pipeline inspector (PI). Remember that the CGA views the pipeline industry as a set of high-level communicating stakeholders and doesn't claim to mandate requirements further down the personnel chain.

Stakeholder involvement

Stakeholder involvement in each practice statement is denoted by an icon, or icons, placed alongside the statement title. Figure 13.2 shows these icons. Where duties of individual stakeholders differ, then these are usually explained in the text of the practice description. Central to most stakeholder involvement is the role of the **one-call centre**, explained in chapter 3 of the document. This is the pivotal resource of the whole best practice 13.0 document, acting as a central communication for underground utility owners and potential excavators. The idea (see Figure 13.3) is that the existence of all underground utilities (pipelines, electrical, services etc.) is identified, recorded and mapped in the 'one-call' database. Modern GPS systems have made this a realistic objective. Anyone who wants to start excavating logs a request (typically 72 hours before intending to start work) with the one-call request line (call 811). If there is no conflict with existing utilities then a permit to dig can be issued. What sounds a simple objective becomes more complicated in practice. Provisions have to be in place to

FIG 13.2
The Pipeline Industry
– The CGA stakeholder group icons –

- operate the one-call system *regionally* across the USA without overlap or dead spots (it's a big country)
- record and administer all enquiry tickets, again without omission or duplication
- handle conflicts between existing installations and proposed new excavations; this requires a system of *location requests* so that routes can be accurately identified
- handle *multiple location requests*: by their nature, pipeline excavations extend over long routes – it is not as simple as planning for an excavation in a single location.

FIG 13.3
CGA Best Practice 13.0
– The one-call centre system –

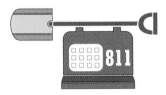

CALL BEFORE YOU DIG (24/7)

- The objective is to avoid accidents to existing underground utilities, particularly pipelines

- It's a legal requirement to call before you dig (72 hours before)

- The one-call system has a governing body including all stakeholders

The USA has regional one-call centres with full national coverage

Once pipelines are established then a marking system is used so that everyone knows where they are. Figure 13.4 shows the idea.

13.2 INGAA pipeline crossing guidelines

Introduction: gas pipeline crossings

According to API's published body of knowledge (BoK), pipeline construction safety accounts for 25% of the API 1169 exam questions. These safety requirements are distributed over more than half the documents in the BoK and reflect the potentially dangerous nature of pipelines and their construction projects. Excavation of a new pipeline trench that crosses the route of an existing gas pipeline is one of the

FIG 13.4
CGA Best practice 13.0
– Pipeline locating and marking –

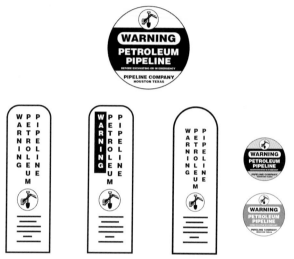

Markers located near roads, railroads, fences and along pipeline right-of-way

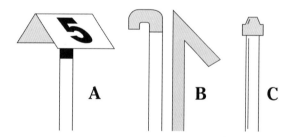

A. Marker for pipeline patrol plane
B. Pipeline casing vents
C. Test station

more difficult operations. Breaching the integrity of an operating gas pipeline by digging equipment produces a serious explosion risk. Accidents like this have happened in several countries with devastating consequences, so steps must be taken to minimise the chances of them occurring.

The INGAA crossing guide

The INGAA *Guidance Document for Construction – Natural Gas Pipeline Crossing Guidelines* is included in the API 1169 BoK code list. This short document (15 pages) covers procedures to be followed when planning and executing the crossing of an existing pipeline route by a new one. Only section II of the document, covering 12 or 13 definitions, is included in the BoK. These are mainly to do with the names and sizes of land areas around the existing pipeline that will be subject to intrusion by excavation for the new crossing pipeline. Section III of the document, giving guidelines on activities to do or not to do in these areas is *not* included in the BoK.

Figure 13.5 shows the main definitions of pipeline crossing land areas divided as follows.

- **Excavation tolerance zone** – a 2 foot wide zone surrounding the existing pipeline in which excavation must be done by non-mechanical means (hand-digging etc.). This avoids the risk of puncturing the pipeline by blasting or mechanical excavators.
- **Active excavation area** – the area within 25 feet of the centreline of the existing pipeline. No excavation may be done in this area until the position of the existing pipeline has been accurately marked.
- **Encroachment area** – the larger area within 50 feet of the centreline of the existing pipeline. It is basically the area in which activities relating to the new pipeline construction have to be closely controlled, under agreement between all the parties involved.

The underlying principle of the INGAA guidelines is that written crossing agreements must be in force between parties responsible for both the existing pipeline and the proposed new one to avoid any misunderstandings. Model crossing agreement forms are suggested.

FIG 13.5
Pipeline crossing terminology*

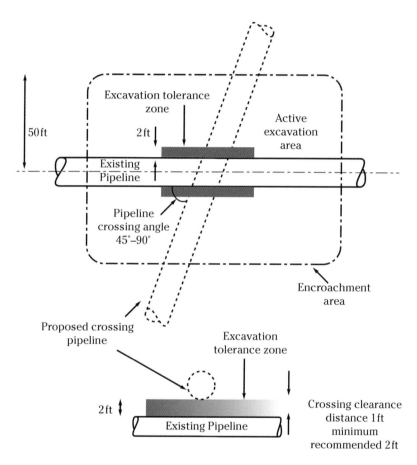

Ref INGAA document: Natural Gas Pipeline Crossing Guidelines
*Note: Only the definitions (Section II) are in the RP 1169 BoK

13.3 Sample questions

CGA best practice document 13.0 (all of it) is included in the main closed-book part of the API 1169 BoK. It does not, however, form a particularly suitable source of PI exam questions due to the length and complexity of its content (100+ pages) and its somewhat arms-length relevance to the work of the PI. Chapters 2–6 of the document contain information that is of benefit to the PI to know, whereas detailed arrangements for community-type objectives and procedures seem of little relevance.

Picking out the most important points of relevance to the PI is not too difficult. They are as follows.

- Pipeline **construction activities** must follow the requirements of the one-call system – a good PI verification activity.
- **Excavation permits** must be in place before digging commences.
- **Pipeline markers** must be set in place after backfilling and reinstatement. Similarly, existing ones must not be removed.
- **Reporting back** to the one-call centre must be done if additional underground services are discovered.

Now try question set 13.1 covering selected parts of the CGA best practice document. You will need to do this open-book to gain the benefit of becoming familiar with the document's content.

Regarding the INGAA crossing guidelines, the terms and definitions used in this document make valid examination questions. Terms and definitions, and definitive information points such as distances and sizes, always make questions easy to formulate. The names and distances are not that difficult to remember for closed-book questions. It's also worth looking at section III of the INGAA guide to see the restrictions that are imposed within the different areas. This is strictly not in the BoK, but will help you remember them by putting them in context.

To help you gain familiarity with the INGAA crossing guide, have a go at question set 13.2.

Question set 13.1: CGA best practice document (13.0)

Q1. CGA best practice manual: reporting and evaluation

From a national data perspective, CGA stakeholders recognise which of the following tools as the most beneficial sources currently available for data regarding damages, near misses and incidents?

(a) DIRT tool ☐
(b) RBI tool ☐
(c) CBYD (call before you dig) tool ☐
(d) PDD tool ☐

Q2. CGA best practice manual

Under the CGA best practice manual, the national 'one-call centre' (call before you dig) number is

(a) 811 ☐
(b) 118 ☐
(c) 611 ☐
(d) 911 ☐

Q3. CGA best practice manual: pre-excavation

According to CGA best practice manual, unless specified otherwise in local/state law, companies wishing to start excavating to install new pipeline facilities should call the 'one-call centre' how long before actually beginning the excavation?

(a) Minimum 30 days ☐
(b) Maximum 10 days ☐
(c) 2–10 days ☐
(d) 1–7 days ☐

Q4. CGA best practice manual: compliance

Under CGA best practice manual rules, some states may issue what type of notice in advance of a Notice of Violation (NOV) alleging that there has been a non-compliance with the one-call centre or related requirements and a type 1 structured review process is in force?

(a) NOAV ☐
(b) NODV ☐
(c) SNOV ☐
(d) NOPV ☐

Q5. CGA best practice manual: definition

In accordance with the definition in the CGA manual, which of the following engineers are involved in identifying underground utility information needed for excavation plans and managing that level of information during the development of a pipeline/facility project?

(a) Project civil engineer ☐
(b) One-call engineer (PCE) ☐
(c) Subsurface utility engineer (SUE) ☐
(d) Ground facilities engineer (GFE) ☐

Q6. CGA best practice manual: stakeholders

The CGA best practice manual indicates the relevance of all its best practice statements and descriptions as they apply to facility owners, excavators, project owners, designers and

(a) Manufacturers ☐
(b) Locators ☐
(c) Legal policy makers (e.g. state jurisdiction) ☐
(d) The general public ☐

Q7. CGA best practice manual: one-call centre: disaster recovery

According to CGA best practice manual, the one-call centre disaster recovery plan is to enable the one-call centre to

(a) Deal with disasters affecting existing pipeline facilities ☐
(b) Predict the effect of a disaster on existing pipeline facilities ☐
(c) Function properly if a disaster affects the centre itself ☐
(d) All of the above ☐

Q8. CGA best practice manual: one-call centre

The one-call centre has a documented and proactive programme covering communication, public awareness, damage awareness and

(a) Technical standards availability and access ☐
(b) Education ☐
(c) Certification of excavation plant ☐
(d) Soil sampling ☐

Q9. CGA best practice manual: locating and marking

If a person given the job of locating existing underground facilities finds errors or omissions in the documentary records they should

(a) Notify the owner/operator through the one-call centre ☐
(b) Notify the owner/operator directly ☐
(c) Amend the records themselves ☐
(d) Stop their work until the situation is resolved ☐

Q10. CGA best practice manual: compliance

According to the CGA, non-compliance in most states of the USA generally results in what types of penalties?

(a) Mandatory criminal prosecution ☐
(b) Mandatory civil penalties ☐
(c) Possible education as an alternative to civil penalties ☐
(d) There are no penalties, except in USA east coast jurisdictions ☐

Q11. CGA best practice manual

The CGA best practice manual is a document primarily involved in specifying well-organised

(a) Protection and preservation of the environment ☐
(b) Management of individual stakeholders ☐
(c) Communication between stakeholders ☐
(d) Construction and testing standards for pipelines ☐

Q12. CGA best practice manual: definitions

CGA defines a ticket number as a unique number allocated to

(a) Any incident report involving a safety issue ☐
(b) A pipeline marker or stake providing a cross reference with ROW plans ☐
(c) An enquiry to a 'one-call centre' ☐
(d) A plat ☐

Question set 13.2: INGAA pipeline crossing guidelines

Q1

Under INGAA pipeline crossing definitions, the rights of way (ROW) are usually established through a written document known as an easement, which confirms that

(a) Communication procedures shall be followed ☐
(b) Legal ownership of each area or asset has been established ☐
(c) Each party has a right to be there ☐
(d) Prior permission must be obtained from the landowner before any other party enters the ROW ☐

Q2

Under INGAA guidelines, the minimum allowable angle for a new gas pipeline to cross the path of an existing gas pipeline is

(a) 90° ☐
(b) 60° ☐
(c) 45° ☐
(d) 25° ☐

Q3

The INGAA 'gas pipeline crossing guidelines' included in the API 1169 Body of Knowledge cover situations in which

(a) New gas pipelines cross a river, waterway or some other natural landscape feature ☐
(b) New gas pipelines cross a highway or railroad ☐
(c) New gas pipelines physically cross the path of existing gas pipelines but do not join them ☐
(d) A new gas pipeline is to be joined to an existing gas pipeline by tee or branch connection ☐

Q4

The INGAA foundation recognises that the construction of pipeline crossings occurs on a regular basis. The development of these guidelines are predominantly about

(a) Safety and communication ☐
(b) Compatibility of pipeline fluids ☐
(c) Pipelines crossing the route of telecommunication equipment ☐
(d) All of the above ☐

Q5

When the route of a new gas pipeline is to cross the route of an existing gas pipeline, the crossing construction is defined as being ended when

(a) All soil has been replaced in the active excavation area ☐
(b) The ROW has been restored but plant and personnel may still be on site ☐
(c) All soil has been replaced in the encroachment area ☐
(d) All construction personnel and plant have been removed from the site and the site boundary/fences have been removed ☐

Q6

Who would be considered as the owner under the INGAA definitions, when a crossing agreement is necessary between the crossing company for the new pipeline and the 'owner'?

(a) The owner of the existing pipeline facility ☐
(b) The owner of the new pipeline facility ☐
(c) The landowner around the existing pipeline ☐
(d) The landowner around the new pipeline ☐

Q7

Crossing clearance distance is the vertical separation between the existing pipeline facilities and where the new pipeline is being installed. Under INGAA guidelines, the minimum vertical separation distance for pipeline crossings is

(a) 3 × pipeline diameter ☐
(b) 24 inches ☐
(c) 36 inches ☐
(d) 12 inches ☐

Q8

Under INGAA pipeline crossing definitions, a pipeline ROW is established through a written document called

(a) Easement ☐
(b) Casement ☐
(c) Boundary rights ☐
(d) Tolerance zone inventory ☐

Q9

When installing a gas pipeline crossing to INGAA guidelines, mechanical excavators should not be used to dig soil

(a) Within the 'due diligence corridor' ☐
(b) Within 24 inches of the existing pipeline ☐
(c) Within 48 inches of the existing pipeline ☐
(d) Anywhere within an established ROW without specific agreement of the landowner ☐

Q10

The due diligence corridor described in the INGAA guidelines is defined as a corridor

(a) The same as the survey corridor ☐
(b) The survey width typically used for biological surveys ☐
(c) Equal to the survey corridor, plus 50 feet on either side ☐
(d) Equal to the survey corridor, plus 15 feet on either side ☐

Chapter 14

Inspector health and safety responsibilities and the BoK

Like it or not, 25% of the questions in the API 1169 PI exam will be about good old health and safety (H&S). Like QA, H&S is one of those subjects about which much is written, even more is said, and it sits squarely as a requirement on all projects large or small. As a participant in any project you are expected to understand (and spread the word) about its priority over most things, and the advantages it brings. After all, we all want to go home safely at night.

We saw in Chapter 5 of this book how the role of the RP 1169 PI is extended outside the purely technical compliance role to encompass environmental protection and H&S requirements. Statistically (and actually), construction sites are much more dangerous than most manufacturing works. They contain more, less predictable hazards, host a shifting sands of contractors and sub-contractors and operate in a site environment that is inherently less stable than a static manufacturing shop. Construction itself involves the dangerous activities of blasting, trenching, pipe laying and backfilling, with waterbodies and hostile environments thrown in for good measure.

14.1 RP 1169's allocation of responsibilities

The PI's H&S responsibilities start from the list of points (nearly 60 of them) given in the API 1169 exam body of knowledge (BoK). It's a formidable list but is easily subdivided into what it sees as the main H&S-related activities, i.e.

- general safety of personnel (e.g. PPE)
- confined spaces
- working at height
- excavation (including blasting)

- welding
- NDE
- pipeline pressure testing
- coatings.

Once you accept the role of the PI in their implementation, these all make sense, representing the H&S risks you will find on a real pipeline construction site.

RP 1169 raises the PI's H&S role in the single paragraph of section 4.11, entitled *Safety*. It then expands on this in chapter 5 *Personnel and pipeline safety requirements*. This contains a breakdown of H&S areas aligned broadly with those covered in the exam BoK. The H&S areas spread over several pages, but all follow the same general pattern – that the PI is expected to have an appreciation of the H&S risks, but not be an expert in the subject. Figure 14.1 shows the breakdown. Terms that are typically used are

- the inspector *should have general knowledge of*...
- the inspector *should know*...
- the inspector *should evaluate*...

Practically, these all mean much the same, reinforcing the view that the PI acts as a verifier and evaluator on site, not the main implementer of the contractor's project H&S management system.

RP 1169 does introduce some pipeline-project-specific aspects of H&S that add to the general requirements that you'd find in factory-based manufacturing. These are

- the requirement for a specific Job Safety Assessment (JSA) for the site work
- permit requirements for rock blasting and trench excavation (a good idea)
- the 'one-call' communication procedure to ensure it is safe to start excavating a trench
- procedures for new gas pipelines crossing existing ones
- atmospheric testing, for hazardous atmospheres in confined spaces.

Only brief details of these are given in RP 1169 itself, specific requirements being contained in other documents in the API 1169 exam BoK list.

Inspector health and safety responsibilities and the BoK 163

FIG 14.1
Summary: The H&S subjects of RP 1169 chapter 5

5.1   Scope
5.2   Job Safety Assessment (JSA)* (specific to pipeline construction)
5.3   Personal Protective Equipment (PPE)
5.4   Loss prevention (I think it means avoiding damaging things)
5.5   Radiation protection
5.6   Site security
5.7   Work permits* (don't just start digging)
5.8   Rigging and lifting
5.9   Energy isolations
5.10 Excavating/trenching safety* (blasting and excavating are not the safest activities)
5.11 Confined spaces
5.12 Atmospheric testing
5.13 Respiratory aids
5.14 Fall protection
5.15 Scaffolding and ladders
5.16 Tools and equipment* (mechanical excavators rather than hand tools)
5.17 Start-up reviews* (gas pipelines = live and dangerous)
5.18 Regulatory inspections* (expect lots, particularly for gas pipelines)
5.19 Vehicle operation* (4×4, 6×6 and 18-wheeler specials)

* These are pipeline-specific activities. The others are fairly standard H&S activities common to most industries

---

H&S: specific BoK regulations

The API 1169 BoK contains two documents in its closed-book question section that relate specifically to H&S practices. These are

- CS-S-9 *Pressure Testing Safety Guidelines* – we looked at the content of this in Chapter 11 of this book
- ANSI Z49.1 *Safety in Welding and Cutting* – technical coverage of safety aspects of site welding of pipeline sections, rather than a statutory regulation, as such.

The remainder are in the open-book questions section of the BoK and consist of statutory codes and regulations. The main ones are the Occupational Safety and Health Administration (OSHA) documents

- OSHA 29 CFR 1910 *Occupational Safety and Health Standards*
- OSHA 29 CFR 1926 *Safety and Health Regulation for Construction.*

Note that not all of the content of the OSHA documents is included in the API 1169 BoK (see Figure 5.1 for details). Equivalent regulations for Canada are the Canada Occupational Health and Safety Regulations (COHS). A couple of the other US Code of Federal Regulations (CFR) documents also contain some specific H&S requirements for pipelines. 49 CFR 192 and 49 CFR 195 cover gas and liquid pipelines, respectively, but only selected sections are included in the API 1169 exam BoK.

14.2 H&S exam questions: treasure hunt

On the face of it, there are an awful lot of pages of H&S-related material in the BoK documents from which to select 25 questions from the 100 in the API 1169 exam. RP 1169 contains multiple cross-references to bits of other codes, distributed among paragraphs of requirements for this and that. Take the documents all together and you are probably looking at one lonely question per 10–20 pages of closely spaced text. Looks difficult to know what to learn. To solve this paradox, it's time for a bit of a lesson in *selectivity*. Let's go on a little treasure hunt...

Treasure hunt

Here's how the treasure hunt for exam questions works. We know there are 25 questions on H&S subjects in the API 1169 exam. Amalgamating the itemised points in the exam BoK and chapter 5 of RP 1169 gives about 19 identifiable sub-topics, once you've eliminated any duplication and a few topics of only minor interest. That makes just over one question per topic, using the fair guess that's how an exam question-setter would go about making the question bank as representative of the subject as possible.

Finding the treasure if still difficult. If you look for the obvious question subjects in chapter 5 of RP 1169, you won't find many hard facts: it's mainly lists of things that the PI should know. Lists of generalities like this are notoriously difficult to formulate closed-book exam questions from, and RP 1169 is in the closed-book part of the exam BoK. The alternative approach, that of selecting the majority of H&S questions from the OSHA open-book regulations, is also a problematic route for the exam-setters. In the examination, the source of an OSHA-based question is notified to candidates by including the words 'according to OSHA' in the question. As the subject of a H&S question is clear to see (e.g. welding safety, working at height, confined spaces or whatever), then the answers are easy to find by just turning to the relevant section of the regulation as referenced in the index. The non-technical nature of H&S information means that multi-choice questions need to be fairly direct, with little room for interpretation or reasoning.

The end result is that, with a bit of concentration, it is not difficult for exam candidates to get all of these types of questions correct. Good news for candidates, perhaps, but not for the overall objectives of the examination. Historically, API ICP examinations have centred on closed-book questions being the real test of a candidate's power of reasoning and understanding the subject, so this just doesn't fit.

An easy solution (the type exam question-setters like) lies in a good old-fashioned compromise. Carefully chosen to satisfy the low energy levels of exam candidates and question-setters alike (it's a committee, after all) it looks like this.

- Assume the 25 H&S questions comprise one from each of the (roughly) 19 sub-topics you can identify from the BoK and RP 1169.
- These 19 questions will be *closed-book*, to check candidates' knowledge, particularly about H&S responsibilities and procedures. Not really exciting stuff, but it should be possible to eke one question from each of the list formats of the BoK and RP 1169 chapter 5.
- As 25 – 19 = 6, the final six questions are chosen as open-book from the collection of OSHA and other H&S documents listed in the BoK.
- Fancy a percentage bet? I would choose these six questions from the specific clauses of the OSHA regulation that are cited in the various paragraphs of RP 1169 chapter 5. There are more than six of these, that's why it's a percentage bet.

Figure 14.1 summarises the breakdown of H&S topics in RP 1169. Presented like this, you should find it easier to see the scope, rather than when they are spread over the various pages of the document itself.

Remembering the rules of the exam question treasure hunt (19 closed-book questions from the BoK/RP 1169 and six from the clauses of OSHA documents cited in RP 1169 chapter 5), all you have to do now is to collect up your documents and look for the treasure, exactly the way the exam question-setter would do. Alternatively, you could rely on the ready-made treasure map that is question set 14.1, which you will find at the end of this chapter. Try the full set of 25 questions, then read the answers and their explanations. You might find some useful insights there... who knows?

14.3 H&S treasure hunt questions: question set 14.1

If you understood all the stuff about the H&S exam question treasure hunt above, then here's the question set you need. At a full-length 25 questions it's a bit too long to comfortably finish quickly. Aim at spending about 90 minutes to try the questions, then follow through the correct answers and the logic that goes with them. That's just under 4 minutes per question – fairly typical for an API-style open-book question. Alternatively, of course you could spend your 90 valuable minutes doing other things; you won't find it difficult to find something marginally more interesting. It all depends on whether you are really interested in finding the exam question treasure or not, and whether you believe it is there and not somewhere else. That's the problem with isolated islands in the Caribbean, isn't it? You just never quite know.

Anyway, the question set comprises 19 closed-book style and six open-book style questions, but feel free to try them all open-book. To make it better for learning you will find the question set first, so you can try them out, with the answers and explanations following. Try them now.

Question set 14.1: RP 1169 and OSHA

Q1. RP 1169: scope

The role of the pipework inspector regarding personnel and general pipeline safety requirements is to

(a) Report any unacceptable practices ☐
(b) Report unacceptable safety practices only ☐
(c) Prevent dangerous procedures being used on site ☐
(d) All of the above ☐

Q2. RP 1169: job safety analysis

What is the role of the PI regarding on-site job safety analysis (JSA)?

(a) They should comment on its content ☐
(b) They should pass the information on to the owner/operator ☐
(c) They should only implement its findings ☐
(d) They should participate ☐

Q3. RP 1169: PPE

Approved head protection (hard hats) should be worn when injury could occur from impact or

(a) Chemical drips ☐
(b) Electric shock ☐
(c) Incorrect recognition of site personnel ☐
(d) Poor visibility ☐

Q4. RP 1169: loss prevention systems

PIs

(a) Evaluate the success of daily safety meetings ☐
(b) Should only attend daily safety meetings if invited by the contractor ☐
(c) Should be capable of organising and conducting daily safety meetings ☐
(d) Need only be capable of understanding and evaluating the proceedings of daily safety meetings ☐

Q5. RP 1169: radiological protection

For NDE personnel handling radioactive sources, the inspector

(a) Shall record the identity of each technician handling the equipment ☐
(b) Shall 'sign off' each technician's Job Safety Assessment ☐
(c) Should be aware of licensing and certification arrangements ☐
(d) Shall countersign the Permit to Work (PTW) ☐

Q6. RP 1169: job site and facility security

On construction sites the PI should be familiar with the

(a) 'Professional Security' publication registration of approved contractors ☐
(b) Handbook RE-5 Building and Site Security Requirements ☐
(c) Safe use of site razor wire regulations; 2010 ☐
(d) Manual of Uniform Traffic Control Devices ☐

Q7. RP 1169: work permits

The requirement for work permits on a pipeline construction site is set by

(a) Owner/operator requirements ☐
(b) Inspector requirements ☐
(c) Local jurisdiction ☐
(d) OSHA inspector ☐

Q8. RP 1169: rigging and lifting safety

According to OSHA, for lifting equipment in use on a construction site, PIs do not require general knowledge of

(a) Hand signals ☐
(b) Equipment labelling ☐
(c) Equipment design factors ☐
(d) Extension and boom height limits ☐

Q9. RP 1169: isolation of energy sources

Exposure to energy or electrical energy may be prevented by

(a) LOTO ☐
(b) IDLH ☐
(c) MOC ☐
(d) The correct QA procedures ☐

Q10. RP 1169: excavation, trenching and boring safety

Who marks the location of existing facilities under the *one-call* system?

(a) One-call locator ☐
(b) One-call responder ☐
(c) The owner-operator of the existing facilities ☐
(d) One-call centre ☐

Q11. RP 1169: confined space entry requirements

In the eventuality of a medical rescue being required from a confined space, the inspector is required to

(a) Help out ☐
(b) Provide a full incident report to the owner/operator ☐
(c) Be trained in confined space rescue ☐
(d) Make sure a plan is in place for this activity ☐

Q12. RP 1169: atmospheric testing

A hazardous atmosphere

(a) Is not covered by PELs ☐
(b) May not be IDLH ☐
(c) Is not below the LEL ☐
(d) Is an O_2 level of 19.5% (the normal amount in the atmosphere is 21%) ☐

Q13. RP 1169: respiratory protection

The terms air-purifying, air-supplied and SCBA apply to types of

(a) Confined space ventilation fans ☐
(b) Respirators ☐
(c) Welding fume extractors ☐
(d) Coating drying methods ☐

Q14. RP 1169: fall prevention

Fall and tripping hazards in new pipeline construction work are primarily due to extensive excavations, constant movement of the work and

(a) Varied terrain features ☐
(b) Absence of handrails and safety ropes ☐
(c) 24-hour working ☐
(d) The difficulty of supervision ☐

Q15. RP 1169: scaffolding and ladders

During a pipeline construction project, scaffolding is normally not used

(a) In conjunction with ladders ☐
(b) For work like valve settings ☐
(c) In trenches ☐
(d) At heights above about 3 feet (nearly 1 metre) ☐

Q16. RP 1169: tools, materials and equipment

According to RP 1169, diesel fuel is what class of liquid?

(a) I ☐
(b) II ☐
(c) III ☐
(d) IV ☐

Q17. RP 1169: commissioning and start-up review

With regard to pipeline owner/operator pre-startup review procedures and checklists, an important aspect for preventing incidents is

(a) Accurate ROW mapping ☐
(b) Regulatory inspections ☐
(c) RBI ☐
(d) MOC ☐

Inspector health and safety responsibilities and the BoK   171

Q18. RP 1169: regulatory agency inspections

With regard to regulatory agency inspections of pipeline construction activities, the PI should

(a) Keep a schedule of planned inspection visits ☐
(b) Refuse them entry until the owner/operator is present ☐
(c) Check their credentials ☐
(d) Not be required to participate at all ☐

Q19. RP 1169: vehicle operation

With regard to site vehicles that are owned by the pipeline contractor, the PI should

(a) Not operate them ☐
(b) Not operate them without supervision ☐
(c) Operate them if required, if licensed ☐
(d) Verify them for compliance with MUTCD ☐

Q20. RP 1169 and OSHA: hazardous materials

According to OSHA, what is the threshold quantity of hydrogen chloride above which it represents a potential for a catastrophic event?

(a) 5 lb ☐
(b) 100 lb ☐
(c) 250 lb ☐
(d) 5000 lb ☐

Q21. RP 1169 and OSHA: hot work permits

According to OSHA, how long shall a hot work permit issued by the employer in a pipeline construction project be kept for?

(a) Until the next daily 'close-out' meeting ☐
(b) Until completion of the hot work activity ☐
(c) Until all welding/cutting/heating etc. equipment has been removed ☐
(d) Until completion of the project ☐

Q22. RP 1169 and OSHA: lifting safety

According to OSHA, how often should lifting slings be inspected by a competent person designated by the employer?

(a) 6-monthly ☐
(b) 12-monthly ☐
(c) Each day, before being used ☐
(d) Before every lift of > 75% SWL ☐

Q23. RP 1169 and OSHA: confined space emergency rescue

According to OSHA, which of these statements are true about employees qualified to making confined space permit rescues?

(a) They shall have a firefighting certificate and experience ☐
(b) They shall have a maximum body weight of 100 kg ☐
(c) Training exercises are required at least every 12 months ☐
(d) Responsibilities shall not be shared between people during a rescue ☐

Q24. RP 1169 and OSHA: excavations

According to OSHA, a trench shield is a structure

(a) To prevent trench cave-in ☐
(b) The same as shoring ☐
(c) Called a 'wales' ☐
(d) To mitigate the effect of a trench cave-in ☐

Q25. RP 1169 and OSHA: soil types

According to OSHA, a trench being excavated in a soil designated as Type A

(a) Is not near heavy traffic ☐
(b) Is inherently unstable ☐
(c) May be in dry rock ☐
(d) May be partially submerged ☐

Now here's the answers to question set 14.1

Q1. RP 1169 (5.1): scope

The role of the pipework inspector regarding personnel and general pipeline safety requirements is to

(a) Report any unacceptable practices ☐
(b) Report unacceptable safety practices only ☐
(c) Prevent dangerous procedures being used on site ☐
(d) All of the above ☐

Answer: (a) RP 1169 (5.1).
Did you fall into the trap of choosing the 'all of the above' answer? It's the last refuge of the hurried or undecided, and those that don't read questions properly. Answer (a) is a straight quote from RP 1169 (5.1). Yes, I know that (c) could be sort of true also, but that's not how it is expressed in the text of RP 1169, so it's not the answer.

Q2. RP 1169 (5.2): job safety analysis

What is the role of the PI regarding on-site job safety analysis (JSA)?

(a) They should comment on its content ☐
(b) They should pass the information on to the owner-operator ☐
(c) They should only implement its findings ☐
(d) They should participate ☐

Answer: (d) Direct wording quote from RP 1169 (5.2).
The PI is expected to participate in the JSA but is not the acknowledged expert on the subject. The JSA could have been amended from those used on similar sites so the PI may not have involvement in every stage of its production.

Q3. RP 1169 (5.3): PPE

Approved head protection (hard hats) should be worn when injury could occur from impact or

(a) Chemical drips ☐
(b) Electric shock ☐
(c) Incorrect recognition of site personnel ☐
(d) Poor visibility ☐

Answer: (b) Direct wording quote from RP 1169 (5.3.3).
Having decided on the answer (b) given by the wording of the

document, the examiner is faced with the task of choosing alternative answer options that are not too outlandish in order to justify their choice of question. You can see from this example that it's not that easy.

Q4. RP 1169 (5.4): loss prevention systems

PIs

(a) Evaluate the success of daily safety meetings ☐
(b) Should only attend daily safety meetings if invited by the contractor ☐
(c) Should be capable of organising and conducting daily safety meetings ☐
(d) Need only be capable of understanding and evaluating the proceedings of daily safety meetings ☐

Answer: (c) Direct wording quote from RP 1169 (5.4.3).
This shows the extended role of the PI as we have discussed in the book. It would not be usual for the PI to be in charge of the meetings (that's the contractor's job). Notice, however, that section 5.4.3 doesn't actually say the PI *does* organise the meeting, just that they *should be capable* of doing so. Word games I'm afraid. Treat this as an example of question-setting technique taking precedence over reality.

Q5. RP 1169 (5.5): radiological protection

For NDE personnel handling radioactive sources, the inspector

(a) Shall record the identity of each technician handling the equipment ☐
(b) Shall 'sign off' each technician's Job Safety Assessment ☐
(c) Should be aware of licensing and certification arrangements ☐
(d) Shall countersign the Permit to Work (PTW) ☐

Answer: (c) Loose quote from RP 1169 (5.5).
This one's a good example of where the exam-setter knows what principle they want to ask about but don't have a particularly good set of test wording to work with. If you look at RP 1169 section 5.5 you can see how the words don't match the answer option (c) exactly but the meaning is clear enough. It's called a *paraphrase*, or, the *question-setter seeing what they would prefer to see*.

Q6. RP 1169 (5.6): job site and facility security

On construction sites the PI should be familiar with the

(a) 'Professional Security' publication registration of approved contractors ☐
(b) Handbook RE-5 Building and Site Security Requirements ☐
(c) Safe use of site razor wire regulations; 2010 ☐
(d) Manual of Uniform Traffic Control Devices ☐

Answer: Yes they're all real (well possibly with the exception of (c)). Either way, you'd be hard pushed to find your way through this one, closed-book. At the expense of wasting some of your precious 90 minutes of H&S treasure question finding-time (I did) it's interesting to look them up, and see what you get.

The Manual of Uniform Traffic Control Devices (known to almost everyone as the MUTCD) *'is a compilation of national standards for all traffic control devices, including road markings, highway signs, and traffic signals'*, it says. And there's more!!

> *2017 brings the 82nd birthday of the MUTCD (82 years; amazing!)... and throughout the year when you see an easy-to-read sign, a bright edge-line marking on a foggy night, the countdown timer at a crosswalk, or a well-placed bike lane, take a moment to reflect on the eighty years plus of progress and innovation that the MUTCD embodies.*

Wow, this has got me reflecting big-time... let's read on...

> *Over the years, the MUTCD has unknowingly become the traveller's best friend and silent companion, guiding us on our way along the streets, bikeways, back roads, and highways. So the next time you hit the pavement, the path, or the pedals, you can be sure that the MUTCD, through dedicated professionals who make complex decisions on what devices to install, will help you get where you want to go safely, efficiently, and comfortably. The MUTCD... it's all about **you**!*

Well done to them I say.

If you are still interested, the answer is (d): RP 1169 (5.6).

Q7. RP 1169 (5.7): work permits

The requirement for work permits on a pipeline construction site is set by

(a) Owner-operator requirements ☐
(b) Inspector requirements ☐
(c) Local jurisdiction ☐
(d) OSHA inspector ☐

Answer: (a) Quote from RP 1169 (5.7.1).
Pretty much a straightforward example reinforcing the principle that the owner-operator is in charge of most procedures that happen on their construction site. PIs needs knowledge of work situations that *require* work permits, but they don't *set the requirements* for them.

Q8. RP 1169 (5.8): rigging and lifting safety

According to OSHA, for lifting equipment in use on a construction site, PIs do not require general knowledge of

(a) Hand signals ☐
(b) Equipment labelling ☐
(c) Equipment design factors ☐
(d) Extension and boom height limits ☐

Answer: (c) Quote from RP 1169 (5.8).
Notwithstanding the straightforward text quote that gives the answer, there are two exam question-setting principles at play here.

- The use of the word *not* in the question, totally changing its sense. The *not* has also been positioned at the beginning of the second line, just where your eyes are most likely to skip over and miss it.
- The concept that the three incorrect answer options **are** mentioned in section 5.8 is primarily what defines that option (c), which **doesn't** appear, is the correct answer. Don't be surprised that there's no qualification of what type of design factors (c) is referring to. It could mean lifting hook or rope factor of safety, which are of interest to inspectors, or it might mean some of the long list of empirical design factors hidden away in the crane gearbox of structure design that would be well outwith the inspector's understanding... it just doesn't say. It also doesn't matter, because the question-setter hasn't thought about it: all they've done is choose a form of words that is

not mentioned in section 5.8 and listed it as an answer option. Simple as that.

Q9. RP 1169 (5.9): isolation of energy sources

Exposure to energy or electrical energy may be prevented by

(a) LOTO ☐
(b) IDLH ☐
(c) MOC ☐
(d) The correct QA procedures ☐

Answer: (a) Quote from RP 1169 (5.9.3) and the abbreviations list (3.2). When question-setters are scratching for meaningful statements to form a question, they can always resort to an acronym. It's a valid technique as API documents usually have a list of terms and acronyms that you are expected to know. As questions, they're a bit weak, with loose or contrived wording to fit the question around recognition (or not) of the correct acronym.

Watch out for answer options such as (d), where you have an (in this case incorrect) acronym formatted with a bit of text, usually containing the word 'correct' or 'approved' to give it an air of legitimacy. It has just been put there to try and divert you into thinking that more words means more chance of it being the correct answer. It's not – QA is to do with the quality compliance of a product or service and has little to do with safety.

Incidentally, LOTO stands for Lock Out Tag Out, IDLH means Immediately Dangerous to Life or Health and MOC is Management of Change – something most companies are quite bad at.

Q10. RP 1169 (5.10): excavation, trenching and boring safety

Who marks the location of existing facilities under the *one-call* system?

(a) One-call locator ☐
(b) One-call responder ☐
(c) The owner-operator of the existing facilities ☐
(d) One-call centre ☐

Answer: (b) from RP 1169 (5.10.2).
Examiners get quite excited when they can pen a question that they would call a *composite*. This is where the correct answer appears in more than one of the codes or documents in the exam BoK. It means that their correct answer is less likely to be argued with, so preserving their

credibility and low excitement threshold, and it forms a good way for candidates to learn also.

That's what you have here; the *one-call system* (basically, a communication system so that people don't go round digging up live gas pipelines) is referenced in RP 1169 section 5.10.2 and explained in detail in the CGA (Common Ground Alliance) best practice document, which is in the exam closed-book section, and is a big document. It's also not immediately obvious from its name exactly what it is... *one call* about what?... so it forms a good choice of exam question.

Incidentally, the CGA best practice guide lives in an, obviously consensual, grammatical world of its own. Look up, if you will, the meaning of the terms *Attribute*, *DIRT field form*, *Orthophoto*, *Plat* and the process of *SUE* . On the committees and organisations side there's *NUCLA*, *CGS*, *PHMSA*, *NTSB*, *FHWA*, *GAUPC*, *GUFPA* and *OSCI* even before you get to *ULCC*, *APWA* and the good old *BP(best practice) committee*. There's always one of those.

Q11. RP 1169 (5.11): confined space entry requirements

In the eventuality of a medical rescue being required from a confined space, the inspector is required to

(a) Help out ☐
(b) Provide a full incident report to the owner/operator ☐
(c) Be trained in confined space rescue ☐
(d) Make sure a plan is in place for this activity ☐

Answer: (d) from RP 1169 (5.11.2).
The inspector is not a rescue professional

Q12. RP 1169 (5.12): atmospheric testing

A hazardous atmosphere

(a) Is not covered by PELs ☐
(b) May not be IDLH ☐
(c) Is not below the LEL ☐
(d) Is an O_2 level of 19.5% (the normal amount in the atmosphere is 21%) ☐

Answer: (c) from RP 1169 (5.12.1).
Innocent though this one appears, you are looking at a triple-barrelled classic. First, see who falls for the *distractor*. Option (d) contains a number percentage (19.5%), distracting some people to think it must be

correct, or why would it be there? Then there is the *attractor*, suggesting that 19.5% O_2 sounds bad as it's less than the normal 21% O_2 in the atmosphere. The definition of hazardous area in section 5.12.1, however, is that must be *below* 19.5% O_2, so (d) is not a hazardous atmosphere.

Then we have the correct answer (c) hidden away in a double negative; *not below* means *higher than*, and higher than the LEL (lower exposure limit) means it's hazardous.

Note the three acronyms in this question, to test if you know what they mean

- PEL is permissible exposure limit
- IDLH is immediately dangerous to life and health
- LEL is lower exposure limit.

You have to watch out for questions like this, particularly in the closed-book part of the exam, where you can be tempted to rush ... and go for (d).

Q13. RP 1169 (5.13): respiratory protection

The terms air-purifying, air-supplied and SCBA apply to types of

(a) Confined space ventilation fans ☐
(b) Respirators ☐
(c) Welding fume extractors ☐
(d) Coating drying methods ☐

Answer: (b) from RP 1169 (5.13).
There are only four lines of text in section 5.13 for a question-setter to squeeze a question from. There's lots of information on respirators in the OSHA documents, however, so don't be surprised if it gets chosen as one of the open-book questions. SCBA is self-contained breathing apparatus – the kind of thing firefighters wear.

Q14. RP 1169 (5.14): fall prevention

Fall and tripping hazards in new pipeline construction work are primarily due to extensive excavations, constant movement of the work and

(a) Varied terrain features ☐
(b) Absence of handrails and safety ropes ☐
(c) 24-hour working ☐
(d) The difficulty of supervision ☐

Answer: (a) from RP 1169 (5.14).
It's the trenches that are the main problem – people have a natural tendency to interrupt getting in or out of the trench to stand on the edge while discussing the most irrelevant of matters.

Q15. RP 1169 (5.15): scaffolding and ladders

During a pipeline construction project, scaffolding is normally not used

(a) In conjunction with ladders ☐
(b) For work like valve settings ☐
(c) In trenches ☐
(d) At heights above about 3 feet (nearly 1 metre) ☐

Answer: (c) from RP 1169 (5.15).
This is just a simple word-for-word quote from the document. It specifically says that scaffolding is mainly used for above-ground work 'like valve settings'.

Q16. RP 1169 (5.16): tools, materials and equipment

According to RP 1169, diesel fuel is what class of liquid?

(a) I ☐
(b) II ☐
(c) III ☐
(d) IV ☐

Answer: (b) from RP 1169 (5.16.2).
Class I is the most dangerous (e.g. gasoline). Diesel is combustible but has a flash point above 100°F (37.8°C) so it is less dangerous. There is no Class IV liquid. This question is not atypical in that it does expect you to remember commonly used facts and figures out of the document for closed-book exam questions. In this case it's a fair technical test to see if candidates understand that stored diesel fuel is not as dangerous as Class 1 gasoline.

Q17. RP 1169 (5.17): commissioning and start-up review

With regard to pipeline owner/operator pre-startup review procedures and checklists, an important aspect for preventing incidents is

(a) Accurate ROW mapping ☐
(b) Regulatory inspections ☐
(c) RBI ☐
(d) MOC ☐

Answer: (d) from RP 1169 (5.17).
MOC (management of change) is cited in section 5.17 as an important issue. Exactly why it is *the* most important is not really made clear. Treat this as a paraphrase example (a poor one at that).

Q18. RP 1169 (5.18): regulatory agency inspections

With regard to regulatory agency inspections of pipeline construction activities, the PI should

(a) Keep a schedule of planned inspection visits ☐
(b) Refuse them entry until the owner/operator is present ☐
(c) Check their credentials ☐
(d) Not be required to participate at all ☐

Answer: (c) from RP 1169 (5.18).
That's what it says.

Q19. RP 1169 (5.19): vehicle operation

With regard to site vehicles that are owned by the pipeline contractor, the PI should

(a) Not operate them ☐
(b) Not operate them without supervision ☐
(c) Operate them if required, if licensed ☐
(d) Verify them for compliance with MUTCD ☐

Answer: (c) from RP 1169 (5.19).
That's what it says (see Q6 for details about the MUTCD).

Q20. RP 1169 (5.2.2e) and OSHA (subpart H): hazardous materials

According to OSHA, what is the threshold quantity of hydrogen chloride above which it represents a potential for a catastrophic event?

(a) 5 lb ☐
(b) 100 lb ☐
(c) 250 lb ☐
(d) 5000 lb ☐

Answer: (d) from OSHA 29 CFR 1910 paragraph 119 (appendix) hazardous materials. This is referenced from RP 1169 (5.2.2e). Note how questions sourced from OSHA should be prefixed with 'according to OSHA', giving you the clue that this is an open-book question.

Q21. RP 1169 (5.7.1) and OSHA 1910.119(k): hot work permits

According to OSHA, how long shall a hot work permit issued by the employer in a pipeline construction project be kept for?

(a) Until the next daily 'close-out' meeting ☐
(b) Until completion of the hot work activity ☐
(c) Until all welding/cutting/heating etc. equipment has been removed ☐
(d) Until completion of the project ☐

Answer: (b) from OSHA 29 CFR 1910 paragraph 119(k). It is referenced from RP 1169 (5.7.1).

Q22. RP 1169 (5.8) and OSHA 1910.184c(14): lifting safety

According to OSHA, how often should lifting slings be inspected by a competent person designated by the employer?

(a) 6-monthly ☐
(b) 12-monthly ☐
(c) Each day, before being used ☐
(d) Before every lift of > 75% SWL ☐

Answer: (c) from OSHA 29 CFR 1910 paragraph 184 c(14). It is referenced from RP 1169 (5.8).

Open-book questions often involve finding the frequency of an inspection activity of some sort, as it is considered information that the inspector should be able to find. To answer this question open-book, the reference to OSHA and lifting should tell you that it is 49 CFR 1910 that you need. Checking the index for *lifting slings* takes you to paragraph 184; it's then a case of looking through until you find list c, *Safe operating practices*, which contains the answer in list item 14.

Q23. RP 1169 (5.11.2) and OSHA 1910.146k: confined space emergency rescue

According to OSHA, which of these statements are true about employees qualified to making confined space permit rescues?

(a) They shall have a firefighting certificate and experience ☐
(b) They shall have a maximum body weight of 100 kg ☐
(c) Training exercises are required at least every 12 months ☐
(d) Responsibilities shall not be shared between people during a rescue ☐

Answer: (c) from OSHA 29 CFR 1910 paragraph 146k, referenced from RP 1169 (5.11.2).
This type of question is awkward and time-consuming. The 'which of these is true?' wording forces you to try to find all four options, the problem being that it can take an awful long time looking through the regulations before you can conclude that something isn't there. Starting with the answer options containing a number is not too bad an idea – they are easier to find than a text statement. In this example, option (c) is the one stated in the regulations in paragraph 146(k). You may find the OSHA regulations are very difficult to navigate. It's easy to find the correct subject section from the index but once there you are faced with a fairly featureless format of i,ii,iii,1,2,3 and a,b,c nested subheadings and it's difficult to keep your bearings when looking through the pages.

Q24. RP 1169 (5.10.1) and OSHA 1926 (subpart P): excavations

According to OSHA, a trench shield is a structure

(a) To prevent trench cave-in ☐
(b) The same as shoring ☐
(c) Called a 'Wales' ☐
(d) To mitigate the effect of a trench cave-in ☐

Answer: (d) from OSHA 29 CFR 1926 subpart P, referenced from RP 1169 (5.10.1).
There are a few specialist terms used for trenching; see them listed in subpart P. Wales are horizontal members of a shoring system parallel to the excavation face whose sides bear against the vertical members of the shoring system.

Q25. RP 1169 (5.10.4) and OSHA 1926 (subpart P): soil types

According to OSHA, a trench being excavated in a soil designated as Type A

(a) Is not near heavy traffic ☐
(b) Is inherently unstable ☐
(c) May be in dry rock ☐
(d) May be partially submerged ☐

Answer: (a) from OSHA 29 CFR 1926 subpart P.
Soil types are classified as stable rock through types A, B, C in order of progressive *instability*. Type A is therefore the most stable soil type but cannot be classed as Type A if it is near heavy traffic. Submerged soil is unstable Type C.

Chapter 15

OSHA health and safety regulations

15.1 The problem with regulations: the unit of truth

Before we start, let me lead you through a little problem. H&S standards are written as *regulations*; legal requirements in the area of their jurisdiction. Legal requirements are written by lawyers who, partly because they want to sound like other lawyers, tend to write in long rambling sentences. If they don't do this, they use more normal-length ones arranged within a structure of nested hierarchies (sections, sub-sections, sub-sub-sections and so on). OSHA 29 CFR 1910, by the way, has *five* such levels so you need to add sub-sub-sub-sections and sub-sub-sub-sub-sections. Very impressive. Lawyers, being lawyers, get quite used to writing and reading this stuff. That is to say they become familiar with bad writing.

Along comes your average technical person and finds all of this next to impossible to read. Reading past a section title, it doesn't take more than a couple of paragraphs before you forget which sub-sub-section you are in and so what the subject is, or was. This, unfortunately is the world we are about to enter in this chapter. The big question is, *why* are H&S regulations written like this? It certainly doesn't make them easy to read and if this book was written in that format you wouldn't get past the first few pages.

I will try and explain. The content of H&S regulations does not, by its nature, suit the long 50+ word sentences containing all the qualifying information that lawyers like. Being almost forced to use shorter sentences means that all the qualifying information lawyers like has to be fitted into multiple levels of hierarchy, so it doesn't escape from the equation, as it were. This is fundamentally what scares lawyers – the fear of not being able to express a proposition (legal term) into a single sentence. They feel that it has to be fitted into a single sentence or there is a danger it will be interpreted incorrectly.

FIG 15.1
The nested hierarchy levels of
OSHA 29 CFR 1910

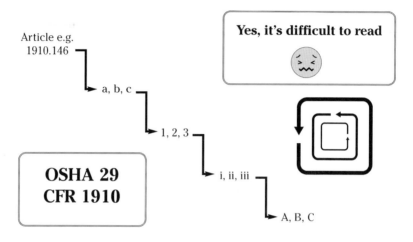

Look for this, and you can see it in action.

For example the requirement;

Before an employee enters a space, the internal atmosphere shall be tested

There it is... in 1910.146 (c) (5) (ii) (C)

Let's express this another way. Lawyers, believe the *unit of truth* is the sentence, when in reality it is the whole of the text of the regulations. Can you now see the reason for the awkward hierarchy structure of OSHA regulations? Coming back to OSHA 29 CFR 1910, you can see this in the layout and Figure 15.1 shows it in all its glory – five levels of nested hierarchy, each believed to be essential to portray the true meaning of the regulations. Now you know the reason, it doesn't, unfortunately, make the thing any easier to read.

If you think about it, this could be good news for open-book exam questions. If the finer technical points of the text, hidden away in this mass of poor writing, are difficult for exam candidates to find, then they will be difficult for the question-setters too. Some may like the idea of a word-puzzle type exam question or two requiring a trail through all the text hierarchies to solve, but most question-setters will default to either

- an easy fact or figure that jumps out at you on the page, without having to bother with the convoluted logic that surrounds it
- more 'general knowledge' style of questions that could possibly be answered closed-book with a bit of experience and common sense.

This last type fits particularly well for questions about H&S subjects. Questions about PPE (personal protective equipment), confined spaces, lifting equipment and the like don't really need to be complicated. From the viewpoint of the pipeline inspector (PI) there is no great technical complexity about these subjects, so there's no point in questions involving complicated reasoning. Straightforward issues of what to do, when to do it, or who is responsible for something, are all that's required. Good news for exam candidates.

15.2 OSHA 29 CFR 1910

Which parts of 29 CFR 1910 are in the exam?

The API 1169 BoK contains just those parts shown in Figure 15.2. Given that 25% of the 100 exam questions are on H&S matters then it is fair to expect a few questions from each subject. On page-count, subpart I covering PPE is by far the largest. A lot of it is, however, taken up with detailed information on breathing respirators, their design, performance and fit-testing – unlikely to be a popular subject for lists of PI exam questions.

Note that 29 CFR 1910 is not the sole H&S content of the exam BoK. ANSI Z49.1 *Welding and Cutting Safety* and 29 CFR 1926 H&S *Regulations for Construction* (see later) also contain information of similar relevance to 29 CFR 1910.

Exam questions

There is little point here in repeating all the technical information points in 29 CFR 1910. It's all written in the regulations themselves and repeated in a multitude of other specialist publications and websites on H&S. The best way to approach this is to have a quick look through the regulations, identifying the sections shown in Figure 15.2 as being included in the BoK. To help you find your way through the confusing hierarchy, have a look back at Figure 15.1. What this figure lacks in excitement it gains by helping clear up the topics included in each of the

FIG 15.2
The parts of 29 CFR 1910 in the exam BoK

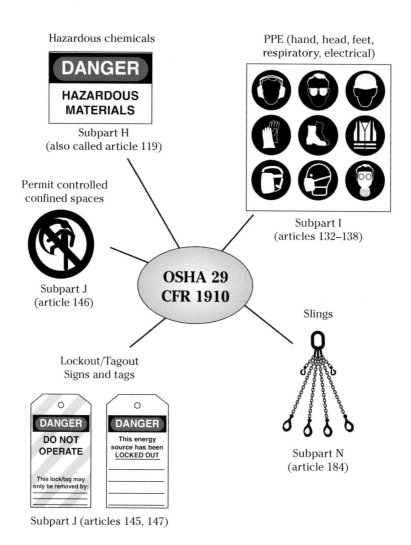

articles of 29 CFR 1910 that are included in the exam BoK. These are listed as a, b, c, d etc. (i.e. the first level of sub-section hierarchy shown in Figure 15.1). Its main purpose is to help guide you to the answers to the sample questions (question set 15.1) at the end of this chapter. Rather than flicking aimlessly through the regulations hoping for the answer to jump out at you, start by deciding what the *subject* of the question is and then see if you can find it listed in the figure. That will lead you straight to the sub-clause required. Remember that you will need to find this among the five levels hierarchy – not that difficult now that you know it is there, and why.

15.3 OSHA 29 CFR 1926

OSHA 29 CFR 1926 addresses the construction-specific H&S issues that fall outside the scope of the general H&S regulations 29 CFR 1910. These regulations are not specific to pipeline construction but do match well with the activities that are found in an overland pipeline construction project. The API 1169 BoK covers eleven main areas of the regulations, divided into various subparts and articles. In general, these are more technically oriented than those covered by 29 CFR 1910, and contain some quite detailed technical requirements, with diagrams to explain the text.

What's in the exam BoK?

Figure 15.3 shows the subjects from these regulations that are covered in the BoK. Of these, the scaffolding and excavation parts are the longest and contain the most instructional detail. Of the two, excavation (subpart P) is the most relevant to the work of the PI. Scaffolding (subpart L) is used in pipeline projects for pipe bridges, above-ground installations etc., but can be considered a lower priority, at least for API 1169 exam purposes.

Remember the role of the PI as H&S inspector?

We discussed this earlier and decided that while the PI does not *lead* the H&S part of a pipeline contract, they do play an active part *in it*. That's why H&S accounts for 25% of the questions in the API 1169 exam. OSHA 29 CFR 1926 (like 29 CFR 1910) is in the open-book question part of the exam, unlike the other H&S standard ANSI Z49.1 *Safety in Welding and Cutting*, which is in the closed-book section.

The structure of 29 CFR 1926 is much the same as that for

FIG 15.3
Regulations for construction OSHA 29 CFR 1926
API 1169 BoK content

OSHA 29 CFR 1910

LEAD
Subpart D
(Article 62)

MOTOR VEHICLES
Subpart O
(Article 601)

FLAMMABLE LIQUIDS
Subpart F
(Article 152)

EXCAVATION
Subpart P
(Articles 650–652)

EYE/FACE
PROTECTION
(Article 102)

CRANES/DERRICKS
Subpart CC
(Article 1417)

MATERIAL
HANDLING/
STORAGE
Subpart H
(Articles 250 and 251)

EXPLOSIVES & BLASTING
Subpart U
(Articles 902 and 914)

SCAFFOLDING
Subpart L
(Article 451)

GENERAL
Subpart C
(Articles 20–35)

FALL PROTECTION
Subpart M
(Articles 500 and 501)

29 CFR 1910 shown in Figure 15.1. The five-level text hierarchy is never going to be easy to navigate through but, nevertheless, 29 CFR 1926 does seem a bit easier to read. The technical subjects are more discrete, more practical and even a bit more interesting. There is more potential for valid open-book technical questions, to add to the usual ones on the role and responsibilities of the PI in H&S matters.

To lead up to the 29 CFR 1926 sample question set at the end of this chapter (question set 15.2) we will look in more detail at the most relevant parts of the document. You can read in more detail what the regulations say about these areas yourself – just follow the topic guide in Figure 15.3.

15.4 Trench excavation: 29 CFR 1926 subpart P

All you need to know about trench excavation (subpart P)

Surprisingly, none of this subpart is about the actual physical activity of excavating a trench (an example is shown in Photo 15.1), it's about checking you have got a safe one once you have excavated it. It starts by listing the bad things that can happen to people in trenches (see Figure 15.4). Most of these are fairly obvious but hazardous atmospheres in particular are a less common but real danger under conditions where heavier-than-air gases are stored on site and may escape into the trench. The regulations make statements on all the hazards shown in Figure 15.4, in practice requiring a risk assessment to be done and prevention/mitigation measure being put in place accordingly (see article 1926.651 for the detail).

Statistically, the greatest danger with trenches is cave-in. Weather conditions and/or unstable soils cause the edges to collapse, burying or injuring personnel working in the trench. Most of subpart P is therefore about avoiding this scenario by using the correct design of trench and having a suitable shoring system in place to prevent danger if collapse does occur. The degree of danger of trench cave-in is directly related to the type of soil the trench is dug in. OSHA specifies (along with ASTM) a system of soil classification, with trench design and shoring restrictions for each type.

Soil classification types

Soils are classified into one of four types: stable rock (even rock is called *soil*), type A, type B or type C, in *decreasing* level of stability. Like metals, soils of broadly similar chemical composition can exhibit

FIG 15.4
Trench safety hazards

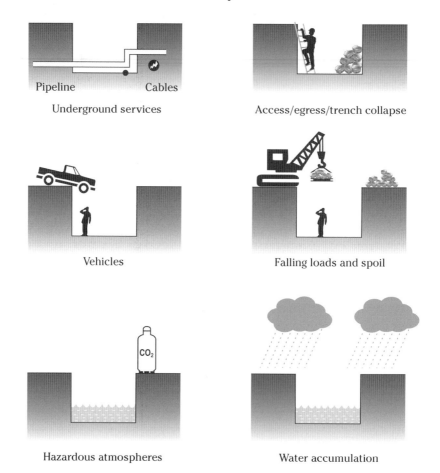

Underground services

Access/egress/trench collapse

Vehicles

Falling loads and spoil

Hazardous atmospheres

Water accumulation

See 29 CFR 1926.651 for the regulations covering these hazards

Photo 15.1 Trench excavation machine (photo courtesy Bigstock)

dramatically different mechanical properties, dependent on their grain structure and trace elements. Water content also has a large effect and, unlike metals, soil microstructure is highly variable and awkward to define. The end result is that a very similar soil composition may exhibit several different levels of stability (A, B or C), making them difficult to classify. Figure 15.5 shows soil terminology used to help describe the differences. Note the information about the 'look and feel' of each type.

Now move to Figure 15.6. This shows the four main types. Compressive strength in tons per square foot (tsf) is a numerical way of determining each classification but note that there are qualifications and exceptions that can move a soil to a less stable classification depending on its location or visible structure. As an example, you can see that, for rock

- stable rock is classed as *stable rock*, the top (A+) stability category

while

- unstable rock that is submerged is type C, on a par with sand, has the least stable category, even though it is still rock.

The type of soil classification chosen is important as it influences the allowable shape of the trench and the arrangement of the shoring system used.

FIG 15.5
Soil terminology
(29 CFR 1926: Subpart P)

Cemented Soil Particles held together by a chemical agent (e.g. $CaCO_3$) Cannot be crushed by finger pressure

Cohesive Soil Fine-grained clay-type soil. Doesn't crumble. Plastic when moist and hard when dry

Fissured Soil Breaks along definite fracture planes with little resistance

Granular Soil E.g. sand. Little or no clay content. Crumbles easily when dry. Cannot be moulded when moist

Moist Soil Looks and feels damp. Can be rolled into a ball without crumbling (cohesive)

Note the definitions are used in the clasification of soil types A, B, C (see Figure 15.6)

FIG 15.6
Excavation: Soil classifications
(29 CFR 1926: Subpart P)

Stable rock

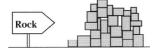

- Can be excavated with vertical sides and remain stable

Type A:
High cohesive soils

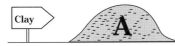

- Compression strength >1.5 tsf (tons/square foot)
- Soil previously undisturbed
- No fissures
- Not near heavy traffic or vibration

Type B:
Low cohesive soils

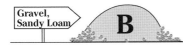

- Compression strength 0.5–1.5 tsf (tons/square foot)
- Type A if fissured or near traffic
- Dry rock if not stable

Type C:
Very low cohesive soils

- Compression strength <0.5 tsf (tons/square foot)
- Sandy, low cohesion, granular
- Submerged soil or unstable rock

Trench sloping and benching

One of the major objectives of 29 CFR 1926 subpart P is to control the allowable shape of trenches. All the information given in the regulations applies for trenches up to a depth of 20 feet, which is deep enough for most pipelines. Trenches that need to be deeper than that have to be treated as a special case and don't fall easily under the scope of these regulations. For depths up to 20 feet, the regulations (1926.652) allow the following possibilities.

- A conservative 'worst-case' sloping limit, assuming a type C (unstable) soil with a 3:2 horizontal:vertical slope ratio (equivalent to 34° from the horizontal). Note how trench slopes are traditionally stated as a horizontal:vertical ratio.

OR

- Follow the slope configuration in 1926.653 appendices A and B. This allows different configurations depending on the soil type.

OR

- Use some other industry-tabulated data. There is no guidance/restriction on what can be used, but a copy must be kept at the jobsite.

OR

- Have the arrangement designed by a registered Professional Engineer (PE), again with a copy kept at the jobsite for reference.

You can see that the regulations accept that the suggested arrangements shown in appendices A and B are not the only acceptable way of doing things, and that other systems may be equally safe. This is a similar approach to technical publications such as ASME codes, which only claim to offer workable, proven solutions, not the only options available. In practice, most pipeline contractors follow the arrangements given in the regulations. Figure 15.7 shows typical examples for type A, B and C soils. Note that these are for long-term trenches; the regulations allow some exemptions for trenches designated as short-term trenches (open for less than 24 hours).

FIG 15.7
Trench sloping and benching limits - Subpart P (Appendix B)

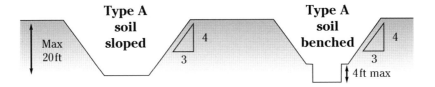

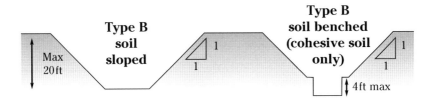

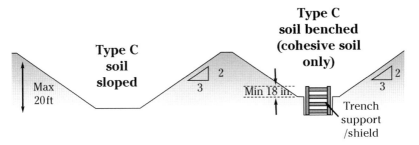

- These are for long-term trenches in non-layered soils
- See subpart P Appendix (B) for other e.g. multi-benched options
- Alternative designs may be acceptable if they follow approved data tables and/or are designed by a Professional Engineer (PE)

FIG 15.8
Timber shoring of trenches

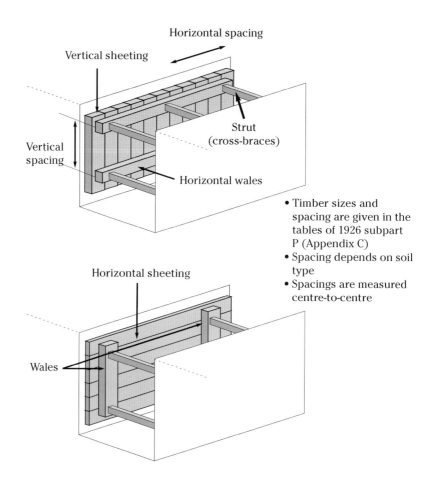

- Timber sizes and spacing are given in the tables of 1926 subpart P (Appendix C)
- Spacing depends on soil type
- Spacings are measured centre-to-centre

Shoring systems

Timber shoring
Shoring is simply some way of supporting trench sides so they don't collapse. Simple wooden shoring consists of thick wooden (timber) planks laid horizontally or vertically along each side of the trench wall braced by supporting members, termed *wales*, perpendicular to them. Cross-braces or struts are then used to brace between the sides, exerting the lateral force that prevents collapse. Figure 15.8 shows the idea. Requirements for timber shoring are given in appendix C of 29 CFR 1926 subpart P. As with trench sloping/benching, various options are given, one of which being to follow the various tables in appendix C for arrangement, dimensions and spacing of the shoring members, depending on soil type. These tables are simple enough to follow and the options are well proven.

Hydraulic shoring
An alternative to timber is sheet metal side-sheets and wales braced by hydraulic cylinders across the trench. This is a quickly assembled and re-usable system suitable for repeated use over long trench distances. Material used can be steel or aluminium alloy and the wales and braces can be 'stacked' to support greater trench depths. Figure 15.9 shows the arrangement. Requirements for dimensions and spacings are given in tables in subpart P appendix D.

Trench shields
A speedier system of trench support is to use a trench shield, as shown in Figure 15.10 and Photo 15.2. These come in pre-assembled sections and are simply lowered into the trench by crane and joined end-to-end. They are in common use as a temporary system for short-duration trenches, repairs etc. For safety, the regulations require that they extend upwards a minimum of 18 inches above the lower edge of the trench slope.

15.5 Blasting and explosives: 29 CFR 1926 subpart U

As part of a federal regulation, subpart U has a surprisingly small scope. It has nothing to do with the *use* of explosives in blasting excavations through rock, nor of the choice, safe handling or procedural aspects of their use on site – these are specialist subjects covered by other regulations. It covers very general aspects of explosives as follows.

FIG 15.9
Aluminium hydraulic shoring of trenches

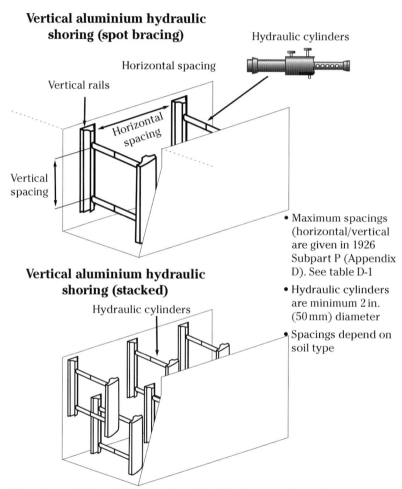

Note that vertical uprights may be replaced by horizontal rails (called wales)

FIG 15.10
A ready-made trench shield

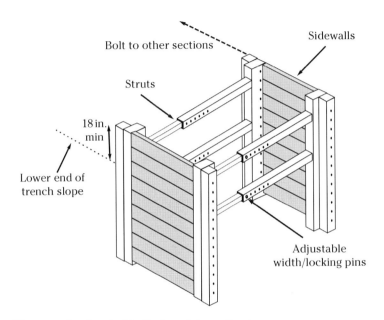

- These are lined up inside the trench in sections
- Minimum vertical projection of 18 in. above the lower edge of the trench slope

1926.900: General principles
1926.901: Blaster qualifications
1926.902: Surface transportation of explosives
1926.914: Definitions.

Of these, only articles 902 and 914 are covered in the API 1169 BoK. Definitions are always included to explain any terms that arise in other articles, so in practice this limits the scope to article 902 *Surface transportation*. Again, the level of detail is very low, comprising general safety points such as vehicle drivers being specially licensed and physically fit, segregation of explosives, detonator items etc. to be transported, and vehicle markings. Figure 15.11 shows the basic points.

Photo 15.2 Trench shield (photo courtesy Bigstock)

API 1169 exam questions?

There isn't much obvious emphasis of explosives/blasting knowledge in the exam BoK. This is not too surprising as it's a specialised subject about which a PI cannot reasonably be expected to have much knowledge. There is no overt requirement for the PI to check personnel qualifications in this field. Regulation 1926.901(c) says that a person involved in blasting shall be *qualified by training, knowledge and experience* without reference to any formal certification structure. Most explosive personnel come from a military background so qualifications and experience are difficult to check. Either way, article 1926.901 *Blasting qualifications* is not included in the API 1169 BoK.

15.6 Operation of cranes and derricks: 29 CFR 1926 subpart CC

Conspiracy against common sense?

It may seem odd, but accidents involving people falling into open trenches are not the biggest source of lost-time accidents on pipeline construction sites. This was even true on a well-documented series of accidents during a 500 mile (800 km) pipeline project (across reasonably

FIG 15.11
Road transport of explosives
29 CFR 1926 (Article 902)

White sign with red letters on all four sides of the vehicle

Red flag with white letters

Specially licenced driver

Vehicles containing explosives:
- Should not be left unattended
- Shall not be taken inside a garage etc for repairs
- Shall have a fire extinguisher
- Shall be correctly marked as shown above

hospitable terrain it has to be said) where a grand total of 50 'fall into trench incidents' were recorded. That's (only) one every 10 miles or so. Most didn't cause serious enquiry, with the overriding root cause of about half of them being that the personnel involved didn't quite expect to find the trench in the place where it was. Even these impressive statistics (people record these things) were herded into second place by accidents involving lifting equipment. You only have to lift something off the ground for a fleeting moment it seems, and it'll find a way to get back there as unpleasantly as possible.

This is a common story. There are more industrial accidents involving lifting equipment than most other sources added together. No surprise then that OSHA and most other regulatory bodies worldwide put lifting equipment and lifting operation near the top of their list for regulatory control. The thrust of most regulations is that duty holders (employers, operators etc.) are required to have written procedures in place to identify all unsafe practices and then act to prevent them. Lifting regulations are among some of the best, having been developed over time, and in the face of enough accidents to help them decide what to protect against. Procedures have been elevated to superstar status in the world of construction projects, with everyone required to agree to them, sign them and comply with their requirements to the letter. Ignore these procedures at your peril.

The problem is that the non-existence of a procedure document is not the reason for most cranes and lifting equipment accidents – it is not *following* a procedure that is actually in place. It may be a question of not having access to a procedure (perhaps because you didn't look for it) or not reading it, thinking that it's someone else's job to comply with it. At the bottom of the list of excuses live the misdemeanours of just simply taking shortcuts or losing concentration on the day.

Procedures don't always work. It makes a nice story, but sheets of paper will not prevent a load swinging, falling or otherwise ending up where it is not supposed to. This is not an argument against procedures, merely to say that they are not the full story. What you need on top of them is *supervision*.

Why supervision is a good thing

If you look at systems of any type, they don't work at all well without the property of *feedback*. This applies whether the system is social, biological, technical or organisational. It really doesn't matter – they all need feedback to function. Safe lifting operations are an organisational system of fairly low complexity compared with many, with the simple remit of getting the lifting task done without mishap. Written procedures (the rules of the system) are there to provide *guidance* to the *inputs* to the system (the physical actions of the crane drivers, banksmen and loader/unloaders involved). The *outputs* are then their *actual* actions that they perform.

Feedback in any system looks at the outputs, comparing them with the desired ones suggested by the rules of the system. It's looking for differences between the desired outputs and those that are actually achieved. There will always be some deviations, large or small, but they will only be found by the activity of *supervision*. What happens to the results of the supervision is equally as important as the supervision itself. Changes need to be made to the input actions to get the outputs back on track. The whole process, simplistically, constitutes *system feedback*. The longer you think about it, the more you should see that this feedback loop can't work without the existence of supervision, which is why it is a good thing.

Supervision is optional, which is why it reveals information

Supervision takes place on several levels. Plant operators themselves act as their own feedback loop; if some activity is not working, they will adopt a different approach. Advice and criticism from others involved in the lifting operation (not in short supply on a construction site) may guide them or not, as the case may be. Project H&S people provide an additional channel of feedback as long as they are physically present on site during the lifting procedure, rather than reviewing procedures in the office. To complete the picture, other involved and not-so-involved project and construction staff all seem to be drawn to watching cranes lifting things, while passing comment among themselves on what is going on.

Most of this supervision involves observations and actions that people are doing *voluntarily*. Their presence may be required by some procedure or other but no-one is telling them, second by second, what exactly to look at. Given their free choice of what to do, information that they do pick up will tend to have a higher value than any observation that is imposed on them. This better quality of information is exactly what the feedback loop needs to work best. Now, let's introduce the PI into this picture.

What should the PI do?

Easy. Check if the lifting operator's feedback loop is working correctly. Although an active participant in the site construction project team, the PI has the advantage of being one (small) step removed from some of the cost and schedule pressures of the project. It's therefore easier to look at the operations without getting caught up with too much of the fine detail. The overall objective is to check whether there is actually anything inhabiting the feedback loop, as there should be. The key point here is that no matter how good a documented lifting operation is, it will be a weak instrument if its implementation is not supervised in a way that works. Indications that a system is not working are

- people making observations, but the procedure document never changes
- the lifting operations procedure that is supposedly being followed is often spoken about but no-one has it with them

- procedure documents seen as a remote thing, to be overridden as necessary by the schedule, cost and logistical realities of life on a pipeline construction site.

These are the cause of lifting accidents, and the source of the need for regulations, which we will look at now.

The regulations

The four pages of subpart CC (also called article 1417) look, for all the world, as if someone has run their finger down some annual review of crane accidents and listed all the causes of accidents found there. Article 1417 consists of a list of no less that 27 of them (*a* to *z*, followed by *aa*), some with a short qualifying statement or explanation. There is no doubt that all of these make good sense, providing a useful checklist for an inspector to use either when reviewing a lifting operating procedure before a lift or as points to observe when the lift itself is in progress. Pipeline lowering-in is a repetitive process so it is important to make sure that lifting procedures are well documented and established across the different working shifts. Most lifting contractors are well aware of the need for approved lifting procedures – it's the actual execution of them where loose practices can creep in.

Figure 15.12 summarises the 27 points of article 1417, all dealing with some aspect of the *lifting operation* rather than the design and inspection of the equipment. The following are worth a special look, due to their conservative, safety-conscious nature.

(d) Prohibition of *mobile phones* from equipment operators. This can be difficult to enforce, but nothing less than a complete ban on them is likely to work.
(f) *Tag-out* of equipment. The regulations give no leeway on timescale for this.
(s) *Brake testing*. These must be tested by the operator *each time* a load > 90% safe working load (SWL) is to be lifted. Procedures should be clear as to whether pipeline lowering-in qualifies for this. Normally it doesn't, but it can become an issue if lower SWL lifting equipment has to be used when normal cranes are being serviced or repaired.
(y) Crane operators must obey a '*Stop*' signal irrespective of who gives it. This differs from some procedures where it is the lifting superintendent (banksman) that provides the definitive instruction to the crane operator.

FIG 15.12
Operation of cranes and derricks in construction 29 CFR 1926 (Article 1417)

The long list of sub-sections

a) Employer responsibility
b) Unavailable operating procedures
c) Accessibility of procedures
d) Diverted attention
e) Leaving equipment unattended
f) Tag-out
g) Starting the engine
h) Storm warnings
i) –
j) Equipment adjustment/repairs
k) Safety devices
l) –
m) Slack rope condition
n) Wind, ice and snow
o) Rated capacity
p) Boom obstruction
q) Sideways dragging
r) Lifting over front
s) Testing the brakes
t) 2 wraps on drum
u) Travelling with a load
v) Rotational swinging
w) Restraint lines
x) Brake adjustment
y) Emergency stop signals
z) Swinging locomotive cranes
aa) Counterweight ballast

These all relate to the *operation* of lifting equipment

All in 29 CFR 1926.147

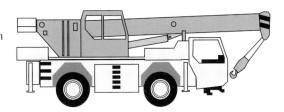

Remember: Lifting equipment is one of the main sources of construction site accidents

There is nothing actually wrong with any of these safety requirements; you just need to make sure that they are understood and implemented properly.

15.7 Scaffolding and fall protection: 29 CFR 1926 subparts L and M

The OSHA 29 CFR 1926 subparts covering scaffolding and fall protection are intended for all construction projects including bridges, buildings, tanks, towers and similar multi-level structures. They are also relevant to pipeline construction, but for most pipelines their application is limited to the few above-ground installations situated along the RoW and protection against falling into open trenches. Occasionally, pipelines may transverse over elevated pipe bridges across land gorges or waterways, so these regulations certainly can't be ignored. Both scaffolding (subpart L) and personnel fall protection (subpart M) are included in the API 1169 exam BoK, probably for completeness rather than common application. On the positive side, many PIs will have background in the NDE or construction industry where scaffolding and fall protection are major H&S subjects and so will have some familiarity with them.

Scaffolding certification: subpart L

Most industrialised countries have a system of certifying scaffolding as safe for use before personnel are allowed to use it. One example is the 'ScaffTag' system. Systems require an authorised 'competent person' supervisor who signs-off coloured tags displayed on ladders and entry routes to the scaffolding, certifying it is safe. Scaffolding is designed and constructed with a large factor of safety (>4) over its certified SWL so problems mainly emanate from bolting the components together. The fact that site scaffolding is always a temporary construction, continually assembled and disassembled, doesn't help.

It is accepted by RP 1169 that the PI is not a scaffolding expert. All the necessary details of level heights, supports, restraints etc. are set out in detail in 29 CFR 1926 subpart L, but this comprises hundreds of individual features and dimensions that a PI could not be reasonably expected to remember. In true API exam style, however, such points still make valid open-book exam questions, as long as the subjects are reasonably easy to track down in the text of the regulations. Overall, this is the case with subparts L and M. They still follow the awkward four/five-level text hierarchy shown in Figure 15.1, but they are short and concise enough for exam candidates to be able to track down the answers to the questions.

Fall protection: subpart M

You can think of this as just an add-on to scaffolding subpart L. Section 1926.451(j) specifies that fall protection is required for personnel in access or working areas greater than 10 feet from a secure lower level. Different arrangements are required for scaffold construction and other activities in which the level is reduced to 6 feet (see 1926.501(b)(b)). Note that this is the same as the depth of trench (6 feet) above which edge barriers are required for fall protection.

Scaffold fall protection devices can consist of various combinations of anchorage, employee harness, deceleration device and lifelines. There are many different proprietary types, certified by various organisations. As with subpart L (scaffolding), exam questions are open-book and so not difficult to answer from this small subpart. The *definitions* section 1926.500(b) contains clear and useful terms and definitions, which make good API-style exam questions. Figure 15.13 shows the sub-section breakdown in subparts L and M. This is the place to start when faced with an exam question on scaffolding or fall protection, before you get lost in the small print.

15.8 Flammable liquids fire risks: 29 CFR 1926 subpart F

As a general safety-related site practice, there is nothing wrong with this as the source of a few API 1169 exam questions. Only a single article (article 152) of subpart F is included in the API 1169 BoK, specifically covering flammable liquids. This can include vehicle fuel and various chemicals used for cleaning, painting, weed killing etc. It covers bulk storage from road tankers and fixed storage tanks through to individual 5–60 gallon drums.

Article 152 looks like a long document at more than 25 pages of technical requirements. You can see the breakdown in Figure 15.14. Look at it a bit more closely and you can see that more than 15 pages of this is taken up by article 152 (L), which covers the design and construction of flammable liquid storage tanks. This is of more relevance to liquid storage and distribution facilities than pipeline construction sites. It is also of little relevance to the role of PIs, who are

FIG 15.13
Scaffolding and fall protection
29 CFR 1926 Subparts L and M

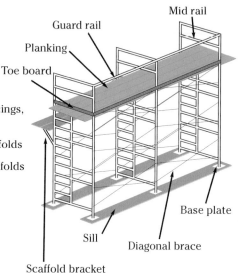

**Scaffolding
(Subpart L Article 45.1)**

a: Load capacity

b: Construction (widths, spacings, restraints, etc.)

c: Criteria for supported scaffolds

d: Criteria for suspended scaffolds

e: Access

f: Use (loadings, defects etc.)

g: Fall protection (>10 feet)

h: Falling object protection

**Fall protection
(Subpart M)**

Article 500 Scope and application

a: Scope

b: Regulations (some good ones in here)

Article 501 Duty to have fall protection

a: Employer's responsibility

b: Unprotected sides and edges

c: Protection from falling objects

more interested in verifying and reporting on operational matters than design features. This leaves the 8–9 pages of items (a) to (g) of article 152 as the main source of information that the PI needs to know.

One feature of any system relating to fire prevention is the regulations of the National Fire Protection Association (NFPA). NFPA has worldwide influence and keeps a tight control on the design and testing of fire systems. Another involved organisation is the Underwriting Laboratory (UL), dealing specifically with storage tanks and receptacles. This helps with the confidence you can have in the design of equipment, leaving more time to concentrate on the operation aspects.

Checking points for the PI

The following points summarise some checking and verification points that fall within the role of the PI. These have all been picked out of the sub-sections (a) to (g) of article 152. It's not an exhaustive list, but does cover many of the main points.

- **Inside and outside storage practices**. There are restrictions on individual liquid container size and how many can be stored together.
- **Identification and marking**. There are several categories (1, 2 and 3) of fluid, based on flash point and the degree of flammability. Correct identification and marking is essential if they are to be stored and used correctly.
- **Handling and spills**. This is where construction site practices can sometimes not be as good as they should be. Construction vehicle fuel is often the biggest problem; in remote regions fuel is stored in bulk on site and spills are not uncommon from filling/emptying couplings. Spill protection bunds etc. are needed both for safety purposes and for environmental protection (see Chapter 16 of this book regarding proximity to waterbodies).
- **Avoidance of ignition sources**. The regulations require close control of ignition sources such as smoking, welding/cutting, electronic equipment and stray electrical currents. These are all good PI monitoring points.
- **Testing of fire protection systems**. Between them, OSHA and NFPA require that fire protection systems (sprinklers, deluge, inert gas flooding etc.) are periodically tested to ensure they work properly. These are common PI witness points.

FIG 15.14
Flammable liquid fire protection
OSHA 29 CFR 1926 Subpart F (Article 152)

Subpart F (Article 152)
Contents

a: General

b: Indoor storage

c: Outside storage

d: Fire control for storage area

e: Dispensing liquids

f: Handling liquids at point of use

g: Service and refuelling areas

h: Scope (<93°C flashpoint)

i: Design/construction of tanks
    Too specialised for PI

j: Pipes, valves, fittings

k: Marine dispensing stations

Construction site points

- Vehicle fuel may be the largest issue for flammable fluids
- Spill prevention is an important monitoring activity
- Fire protection systems should be periodically tested

Fire protection sprinklers

15.9 Materials handling: 29 CFR 1926 subpart H

Materials for pipeline construction

Subpart H of 29 CFR 1926 refers to material transported to site for construction purposes, rather than anything to do with trench excavation or other groundworks. The same regulations would apply to intermediate storage in any delivery depot or warehouse under the control of the pipeline project. Note that the materials involved are assumed to be non-hazardous; if they are hazardous by their chemical composition then additional regulations would apply to cover the particular nature of the hazards involved.

Construction materials on site are generally stored in bagged or stacked form. Subpart H doesn't really cover materials stored in bulk (pile) form. Bagged or stacked materials can comprise

- stone blocks or bricks
- sand and cement
- wood (lumber) for shoring or concrete formwork
- pipe lengths
- reinforcing bars for concrete
- powder or granular chemicals.

Subpart H divides the requirements for handling such materials into two separate articles: article 1926.250 *General requirements for storage* and article 1926.251 *Rigging equipment for material handling*.

Figure 15.15 shows some of the requirements for storage covered by article 1926.250. The main emphasis is on storing the material in such a way that it remains stable and doesn't fall over. Falling material stacks are not an uncommon occurrence on construction sites. Site storage arrangements are, by definition, temporary and so can lack the good organisation and practices of permanent storage yard or warehouse sites. Stack heights and spacing are important points. Brick stacks are limited to a height of 7 feet and the stack has to be tapered back once it reaches 4 feet high. Lumber (wood) piles, being lighter and more stable, can extend to 16–20 feet high depending on whether they are stacked manually or by loading machine.

Requirements for rigging equipment

There is nothing particularly unique about rigging equipment used for materials handling compared with that used for any other type of industrial lifting. The requirements of subpart M article 251 are therefore similar in many respects to those of subpart CC. From a PI's viewpoint the main inspection activities will relate to portable items of equipment, that is

- checking identification of the lifting accessories
- verifying SWLs
- checking their inspection status (they are statutory items that require periodic inspection)
- checking physical condition.

FIG 15.15
Materials handling
29 CFR 1926 Subpart H

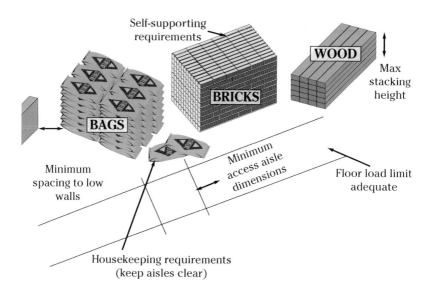

- Article 1926.250 covers general requirements as above
- Article 1926.251 covers rigging equipment for material handling
- See these articles for specific information

Important acceptance criteria for visual inspections of wire slings, alloy steel chains, natural and synthetic rope are identified in separate subsections of 1926.251. These are all safety-related and cover defects such as abnormal wear, broken fibres, stretching, kinking and knots. They make a good inspection checklist. In most construction projects, the RP 1169 PI will not be the only person with an interest in the condition of rigging equipment. Employers may employ an independent 'third party' inspector to inspect them to statutory (jurisdiction) requirements – it all depends on the country and jurisdiction involved.

15.10 Hazardous substances: OSHA 49 CFR 172

Hazardous substances: definition

A substance possessing toxic, reactive, flammable or explosive properties

Pipeline construction projects can have their fair share of hazardous substances hanging around on the site. Most overland projects involve the use of acids and caustics, fuels and oils, bottled gases, paints and cleaning solutions as well as solutions to kill unwanted vegetation, vermin and other pests. For the purpose of the exam BoK, note that the definition of hazardous substances does not include the process fluid (gas, fuel etc.) that will be transported by the pipeline when it is put into operation. This may be hazardous in its own right but is treated differently by codes and regulations.

PI involvement?

The PI is neither a chemical/process engineer nor an expert on the transportation and storage of hazardous substances. There is a network of such specialists in the background behind the decisions to use such materials on a pipeline project. This leaves PIs free to perform their usual function of checking, verifying and reporting that regulations and site procedures are being complied with.

Hazardous substances are well represented in the bunch of regulations that make up the open-book question section of the API 1169 exam BoK. The main general content is in OSHA 29 CFR 1910 (article 119), the set of regulations that deals with general occupational safety and health standards in US industry. This gives a breakdown of 16 topics listed (a) to (p). These are difficult to identify clearly in the multiple sub-hierarchies of the way that this code is written, but they are there, if you look for them, spread over 12 pages.

The other H&S regulations (OSHA 29 CFR 1926) are specifically about construction safety (in all industries). 29 CFR 1926 contains 20–30 pages under the categories of hazardous liquids and materials handling, a lot of which relates specifically to inflammable liquid and the need for safe storage and spill prevention systems.

Hazardous materials table: OSHA 49 CFR 172

OSHA 49 CFR 172 is the main BoK document relating to hazardous substances as they influence the knowledge requirements of the PI. The table is 100+ pages long covering all commonly used chemical substances used in the process industry. The table contains information on the transport of the substances, rather than their process use or hazards. Figure 15.16 shows the format, which is consistent throughout the table. Full explanations of each of the columns are given in the first 11 pages of 49 CFR 172. Note that there are a few columns that contain symbols and abbreviations. These can be a bit confusing at first but easy to understand once you've looked up their meanings in the key. Information is included on sea, rail and air transport. One of the most important columns is the one showing the maximum quantity of a substance that can be transported or stored in one place. There are also some substances for which transport is forbidden for safety reasons. These fall outside 49 CFR 172 and special arrangements need to be made.

Exam questions

Open-book exam questions about 49 CFR 172 are easy to answer once you have identified this document as the source of the question. The table columns are easy to interpret, with the explanations and symbol keys at the front. Any exam question that mentions a specific chemical substance will almost certainly be from the hazardous materials table – none of the other API 1169 BoK documents contains that level of detail. Question set 15.3 gives you some examples to help introduce you to the table columns and symbols. Do these all open-book: there is absolutely no point in trying to guess the answers.

Hazardous substances: PI vigilance

On a real pipeline construction site, a PI has the responsibility to ensure that risks from hazardous substances do not arise from *mistakes*. Mistakes are the most common source of hazardous substance incidents. Familiarity with the existence of 49 CFR 172 and what it contains is therefore practically very useful. Some typical PI verification activities for which it can provide guidance information are

- identifying substances from their shipping code when they arrive at site
- checking deliveries are directed to the correctly allocated storage areas (with spill control, fire prevention etc.)
- verifying that spills are dealt with correctly, in relation to their level of hazard
- ensuring safety signs, vehicle access restrictions etc. are suitable.

FIG 15.16
The 49 CFR 172 hazardous materials table

1	2	3	4	5	6	7	8	9	10
Symbol	Full shipping name	Hazard class	I.D. No	Packing group	Label codes	Special provisions	Packing authorisation	Quantity limits	Ship stowage location
This qualifies the shipping name	Chosen from the international recognised list of chemical substance names	Whether transport is allowed	Linked to shipping name	Type of packing for shipping	Hazard warning label type required	Only special requirements for packing or shipping	Bulk or non bulk packaging	Safe quantity allowed to be transported	Whether it is transported as ship hold or deck cargo etc.

Symbols available are +, A, D, G, I, W

Particularly dangerous substances are forbidden

UN prefix is suitable for shipping

Safe limits for transport by
• Aircraft
• Rail

Categories A to E define on/under deck in various types of ship, cargo/passenger etc.

See pages 1–11 of 49 CFR 172 (Subpart B) for full description of columns 1–10

15.11 Welding and cutting safety: ANSI Z49.1

For some reason the only safety-only document included in the closed-book exam question part of the API 1169 BoK is ANSI Z49.1 *Safety in Welding, Cutting and Allied Processes*. Of the 25% of exam questions covering H&S issues the published BoK makes the odd mention of welding- and cutting-related qualification requirements, permits, dangers of flashing and burning and the need for suitable PPE (as always). It is not treated as a major issue for pipeline construction site work, however, perhaps in the way it would be for fabrication shop manufacture. Notwithstanding this, there is a lot more information in OSHA 29 CFR 1910. Subpart I covers PPE, with a lot of information being about breathing respirators. This is also in the BoK but part of the open-book exam scope, so perhaps a bit less important – the exam is unlikely to contain 25% of its questions about PPE.

What's covered in ANSI Z49.1?

Only four chapters of ANSI Z49.1 are contained in the API 1169 BoK. These are

- chapter 4: Protection of personnel
- chapter 5: Ventilation
- chapter 6: Fire prevention and protection
- chapter 8: Public exhibitions and demonstrations.

These are all quite short chapters, no more than 2–3 pages each, with no illustrations. Chapter 4 confirms the fairly obvious requirements for eyewear, footwear, headgear, protective clothing and respirators. Apart from some detailed information on helmet lenses, there is not much technical detail – that is covered in OSHA 29 CFR 1910. Chapter 5 describes the alternative methods of forced air ventilated welding helmets or local fume ventilation. Chapter 6 provides specific information on fire protection when welding or cutting, such as the use of fireproof materials, installed sprinkler systems and the need for a firewatch during hot work. Chapter 8 (*Public exhibitions and demonstrations*) is a slightly unusual one to include in the PI BoK. It addresses the additional safety requirements necessary when a hot-work process is being demonstrated to an audience, such as at trade shows etc. This doesn't have much to do with pipework construction (not a spectator sport), although could tentatively apply to watching trainees or other construction personnel in close vicinity to site welding.

Links with site construction?

If you read the content of ANSI Z49.1 it is clear that its use was intended mainly for fabrication activities in manufacturing works, plus perhaps weld procedure specifications/procedure qualification records (WPS/PQR) qualification and welder training. It is still applicable to site welding, however. PPE requirements will be the same but the conditions relating to fume ventilation and fire protection will be different on site to those in a factory. Pipeline site welding is done outdoors but in many cases the welding area is practically enclosed by plastic sheet tents so local fumes can still be an issue. Similarly, for welding techniques where manual or semi-automatic weld runs are deposited on the weld joint from inside the pipe, adequate ventilation becomes important.

What about exam questions?

As with all API ICP examinations, candidates are questioned on the *written content* of codes and documents referenced in the exam BoK, not on their general knowledge about what happens in real projects. For this reason, you can expect H&S questions to be drawn directly from ANSI Z49.1 or the two OSHA CFR H&S regulations, depending on whether you feel they apply to cutting or welding of pipelines on a construction site. Don't try to put your own interpretation on such questions – just answer them from the words written in the code or regulation.

Question set 15.4 provides some sample questions drawn from ANSI Z49.1. Before you attempt them

- have a look at the points highlighted in Figures 15.17 to 15.19
- quickly scan through OSHA 29 CFR 1910 subpart I PPE, particularly the section on respirators.

FIG 15.17
Welding Area Fire Protection (ANSI Z49.1)

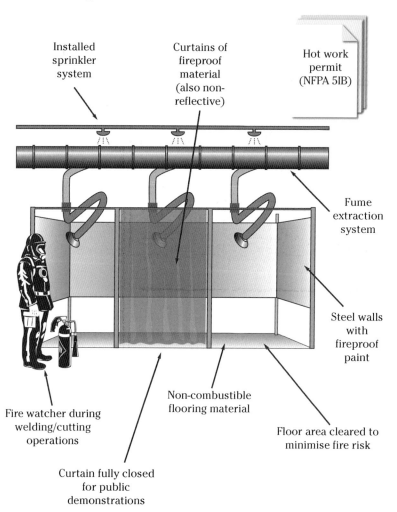

FIG 15.18
Welding booth safety features (ANSI Z49.1)

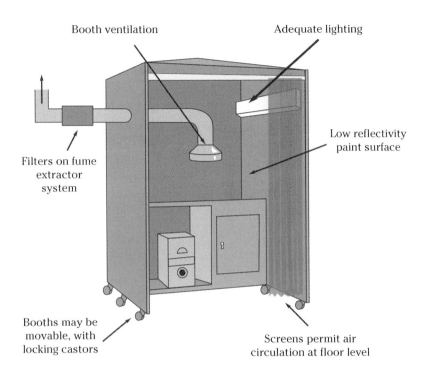

For full details see ANSI Z49.1 *Safety in welding, cutting and allied processes* parts 4,5,6 and 8

FIG 15.19
Welding PPE (ANSI Z49.1)

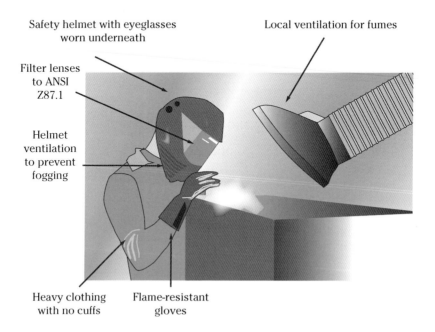

Other safety/PPE points

- Ear protection required for noisy processes e.g. arc gouging
- Respiratory PPE is required if local ventilation is not available
- Ventilation airflow should be across the face of the welder: Z49.1 (E5.3)
- Special precautions required when particularly hazardous metal fumes are involved e.g. lead, copper, nickel, chromium and others. See Z49.1 (5.5)
- See also OSHA 29 CFR 1910 Subpart I. This contains a lot of information on respirator requirements

Question set 15.1: OSHA 29 CFR 1910

Q1. OSHA 29 CFR 1910: hazardous chemicals (appendix C)

According to OSHA 29 CFR 1910, which of the following diagrams depict the process flow details, describing the relationship between equipment and instrumentation, for the use of design and engineering?

(a) Process flow diagram ☐
(b) Block flow diagram ☐
(c) HAZOP diagrams ☐
(d) Piping and instrumentation diagram ☐

Q2. OSHA 29 CFR 1910: lockout/tagout removal

Who shall authorise the removal of a tag attached to an energy isolation method, in accordance with the recommendations provided in OSHA 29 CFR 1910?

(a) Any affected person acting individually ☐
(b) The authorised person's supervisor ☐
(c) All affected persons acting together ☐
(d) The authorised person ☐

Q3. OSHA 29 CFR 1910: lockout/tagout

OSHA 1910 lockout/tagout procedure requirements do not cover the unexpected release of energy from equipment

(a) During servicing on the employer's premises ☐
(b) During servicing at the manufacturer's premises ☐
(c) During maintenance ☐
(d) In normal use following safe procedures ☐

Q4. OSHA 29 CFR 1910: PPE

According to OSHA, the employer shall identify a physician or which of the following detailed below to perform medical evaluations of employees?

(a) QNFT ☐
(b) QLFT ☐
(c) PLHCP ☐
(d) APF or ILDH ☐

Q5. OSHA 29 CFR 1910: hazardous chemicals training

According to OSHA, the employer, in consultation with the employees involved in operating the process, shall determine the appropriate frequency of refresher training. However, refresher training shall be carried out at least

(a) Every 5 years ☐
(b) Every 2 years ☐
(c) Every 3 years ☐
(d) Every 10 years ☐

Q6. OSHA 29 CFR 1910: sling procedures

According to OSHA slinging requirements, which of these is a permitted activity during a slinging operation?

(a) Using a sling shortened by knots ☐
(b) Using slings with padding between them and the load ☐
(c) Imposing shock loads on a sling during the lifting procedure ☐
(d) Using slings without legible identification markings ☐

Q7. OSHA 29 CFR 1910: slings

According to OSHA, what is the maximum allowable number of randomly distributed broken wires in one rope lay of a wire sling before it must be removed immediately from service?

(a) 9 ☐
(b) 4 ☐
(c) 10 ☐
(d) 6 ☐

Q8. OSHA 29 CFR 1910: signs

According to OSHA, the standard background colour of safety instruction signs shall be

(a) Yellow ☐
(b) White ☐
(c) Black ☐
(d) Red ☐

Q9. OSHA 29 CFR 1910: hazardous chemicals

According to OSHA, employers shall certify that they have evaluated compliance (using an audit with the requirements of OSHA 29 CFR 1910 (1910.119)) hazardous chemical requirements at least

(a) Yearly ☐
(b) Every 3 years ☐
(c) Every 10 years ☐
(d) Every 5 years ☐

Q10. OSHA 29 CFR 1910: PPE standards

According to OSHA, standard ANSI Z41-1999 covers what type of PPE?

(a) Safety footwear ☐
(b) Eye guards ☐
(c) Respirators ☐
(d) Safety gloves ☐

Question set 15.2: OSHA 29 CFR 1926

Q1. OSHA CFR 1926: subpart P: soil classifications

Which of the following would not permit a heavy clay soil to be classified as a Type A soil, in accordance with the guidelines provided in OSHA 29 CFR 1926?

(a) If it has a fine-grained structure
(b) Under any circumstances
(c) If it has a cohesive strength > 1.5 ton/sq.ft
(d) If there is road carrying heavy traffic nearby

Q2. OSHA CFR 1926: scaffolding

When may forklift trucks that will not be moved horizontally while the scaffold is occupied be used to support scaffold platforms?

(a) If the forks are only supporting part of the scaffold platform
(b) If the arrangement is just used to bridge a ground-level pipeline
(c) If the forks are supporting the whole of the scaffold platform
(d) Forklifts shall never be used to support scaffold platforms

Q3. OSHA CFR 1926: flammable liquid storage

The storage of containers and flammable liquids outside buildings of not more than 60 gallons each shall not exceed which of the following quantities in any one pile or area?

(a) 600 gallons
(b) 2200 gallons
(c) 1100 gallons
(d) 360 gallons

Q4. OSHA CFR 1926: subpart P: trench excavated safety

When do excavated trenches that are open to the atmosphere and situated in landfill areas need to be checked for oxygen levels?

(a) They don't, as they are open to the atmosphere
(b) Only if they are more than 4 feet deep
(c) Always
(d) Only if they are more than 6 feet deep

Q5. OSHA CFR 1926: subpart P: maximum allowable slopes

According to OSHA (1926 Annex B), steeper excavation side slopes are permitted for 'short term' trenches less than 12 feet deep that will be open for less than 24 hours excavated in

(a) Type A soil and solid stable rock only ☐
(b) Type A soils only ☐
(c) Type A, B and C soils only ☐
(d) All soil types ☐

Q6. OSHA CFR 1926: subpart CC: 'cranes and derricks' repairs

If repairs or adjustments are found to be required for cranes or derricks that remain in use, the operator must inform in writing a person designated by the employee suitable to receive such information and

(a) Also inform the operator following on the next shift ☐
(b) Tagout/lockout the equipment item ☐
(c) Approve the scope of the repairs ☐
(d) Notify all affected employees ☐

Q7. OSHA CFR 1926: subpart CC: 'cranes and derricks' stop signals

When must a crane or derrick operator obey a 'stop' or 'emergency stop signal' when they are in the process of a difficult joint lifting operation, in accordance with the guidelines detailed in OSHA 1926?

(a) It is absolutely at the discretion of the crane operator ☐
(b) Always, even if it comes from an uninvolved person passing by ☐
(c) Only if it comes from any of the persons directly involved within the lift ☐
(d) Only if it's given by the Competent Person supervising the lifting operation ☐

Q8. OSHA CFR 1926: materials handling: sling inspections

Alloy steel chain slings shall be periodically inspected on a regular basis and in no event at intervals greater than once every

(a) 12 months ☐
(b) 6 months ☐
(c) 3 months ☐
(d) 24 months ☐

Q9. OSHA CFR 1926: scaffold fall protection

Fall protection of some type shall be required for every employee on a scaffold where the potential fall could be greater than

(a) 8 feet ☐
(b) 6 feet ☐
(c) 10 feet ☐
(d) 1 foot ☐

Q10. OSHA CFR 1926: lead procedures

When an employee is asked to work in a lead environment with an atmosphere above the permitted exposure limit (PEL), the employee shall

(a) Place all 'contaminated' notices or signs around the work area ☐
(b) Be provided with a respirator ☐
(c) Purchase their own respirator if they want one ☐
(d) Refuse to work in this environment, with or without a respirator ☐

Q11. OSHA CFR 1926: subpart P: excavation definitions

A footing pier hole excavation in which the bottom of the excavation is wider than the cross-section of the excavation above it is called a

(a) Reverse taper hole ☐
(b) Bell bottom hole ☐
(c) Teardrop hole ☐
(d) Trapezoidal hole ☐
☐

Q12. OSHA CFR 1926: materials handling: wire slings

Which of the following statements should be considered as correct regarding the wire rope slings used for material lifting/handling?

(a) Shall have eyes formed by wire rope clips ☐
(b) Shall be used if there is any visible corrosion present ☐
(c) Shall have the protruding ends of strands covered or blunted ☐
(d) Shall have eyes formed by knots ☐

Q13. OSHA CFR 1926: welding safety

Which of the listed materials would necessitate the requirement of LEV and/or respirators due to the dangerous concentrations of nitrogen dioxide, when subjected to inert gas metal arc welding?

(a) All steels ☐
(b) Carbon steels ☐
(c) Stainless steels ☐
(d) Low alloy steels ☐

Q14. OSHA CFR 1926: subpart CC: 'cranes and derricks' procedures

In accordance with OSHA 1926, procedures for the operational controls of cranes must be developed by

(a) The operating personnel who will be using the equipment ☐
(b) The manufacturer ☐
(c) A qualified person ☐
(d) A registered Professional Engineer (PE) familiar with the equipment ☐

Q15. OSHA CFR 1926: definitions

According to OSHA, the 'action required' level of employee exposure to lead or lead-containing products over an 8-hour time-weighted average is

(a) $30\,\mu g/m^3$ air ☐
(b) $40\,\mu g/m^3$ air ☐
(c) $50\,\mu g/m^3$ air ☐
(d) $45\,\mu g/m^3$ air ☐

Question set 15.3: 49 CFR 172 hazardous materials table

Q1

Which one of these is a proper shipping name for a chemical suitable for international transportation detailed within the hazardous chemical table in 49 CFR 172?

(a) Diborane mixtures ☐
(b) Methanol ☐
(c) Activated carbon ☐
(d) Diesel fuel ☐

Q2

According to the 49 CFR 172 hazardous chemical table, the maximum quantity of methyl butyrate that may be transported by passenger rail is?

(a) 50 gallons (in a single container) ☐
(b) 5 litres ☐
(c) 20 litres ☐
(d) 5 kg ☐

Q3

According to the 49 CFR 172 hazardous chemical table, which one of the following chemicals is it forbidden to transport?

(a) Bromosilane ☐
(b) Bromoform ☐
(c) Bromopropyne ☐
(d) Bromotrifluoroethylene ☐

Q4

In 49 CFR 172 hazardous chemical table column 6 showing label codes, a label code 7 signifies a chemical that is

(a) Explosive ☐
(b) Radioactive ☐
(c) Poisonous ☐
(d) Flammable ☐

Q5

A chemical with the abbreviation NA in column 4 (ID number) is?

(a) Not suitable for transportation within the USA ☐
(b) Exempt from the requirements of document 49 CFR 172 ☐
(c) Suitable for all international shipping ☐
(d) Suitable for transportation only within USA and Canada ☐

Question set 15.4: ANSI Z49.1: safety in welding & cutting

Q1. ANSI Z49.1: PPE

Materials which can melt and cause severe burns should not be used as clothing when welding or cutting. Protective clothing treated with flame-resistant chemicals used during welding or cutting operations may lose some of its protective characteristics after

(a) Repeated washing or cleaning ☐
(b) Exposure to sunlight ☐
(c) Exposure to welding arc flashes ☐
(d) Use for too long without washing or cleaning ☐

Q2. ANSI Z49.1: welding screens

Flame-resistant shields or screens are used to protect the workers adjacent to the welding area from heat and spatter. The screens positioned around welding areas should

(a) Have high reflectivity to UV radiation from arc flashes ☐
(b) Extend from floor to ceiling to prevent stray arc flashes ☐
(c) Have gaps above and below them to allow airflow ☐
(d) Not be required if welding operators are using the correct eye, face and body protection ☐

Q3. ANSI Z49.1: PPE

When welders are welding using a 'tilt raise' welding helmet with filter lenses then

(a) Only SAW or Oxyfuel welding processes should be done using this type of welding helmet ☐
(b) Eye protection spectacles shall not be worn under the helmet ☐
(c) Eye protection spectacles shall be worn under the helmet ☐
(d) The helmet shall not be raised until the welder turns to face away from the work ☐

Q4. ANSI Z49.1: fire protection

Hot work capable of initiating fires or explosions requires authorisation

(a) By the local jurisdiction or fire department ☐
(b) Only if there are combustible floors in the area ☐
(c) In the form of a written permit ☐
(d) At least 24 hours before the procedure is scheduled to start ☐

Q5. ANSI Z49.1: public demonstration

When a welding or cutting demonstration is being made to the public at, for example, a trade show site, then

(a) Gas cylinders shall not be filled to more than 50% capacity ☐
(b) Each person in the audience must be provided with protective eyewear, whatever the process and other safety arrangements in force ☐
(c) The local state or county fire department must be present ☐
(d) None of the above are required ☐

Q6. ANSI Z49.1: ventilation

When welding materials that involve fluorine, the fumes can cause dizziness or nausea called 'metal fume fever', so special ventilation is normally required. The same precaution must be taken when specifically welding, cutting or brazing

(a) Corroded components ☐
(b) Galvanised components ☐
(c) Stainless steel components ☐
(d) Any components outside, in a salt-laden marine atmosphere ☐

Q7. ANSI Z49.1: PPE

Welding helmets without goggles underneath will not protect against the effects of

(a) Ultraviolet radiation ☐
(b) Grinding splinters ☐
(c) Fragmenting grinding wheels ☐
(d) Any of the above ☐

Q8. ANSI Z49.1: PPE

Which of the following welding processes would not permit the use of respiratory equipment?

(a) PAW ☐
(b) SMAW ☐
(c) TB (torch brazing) ☐
(d) None of these ☐

Q9. ANSI Z49.1: ventilation

When fumes are produced by a welding process and it is necessary to determine the chemical composition of the fumes by sampling, the samples shall be collected from

(a) The general atmosphere in the welding room or enclosure ☐
(b) Inside the welder's helmet ☐
(c) As near as possible to the weld itself ☐
(d) The area of densest visual smoke around the weld ☐

Q10. ANSI Z49.1: fire protection

How long should a fire watch be maintained when hot work is being carried out and there is a possible fire risk?

(a) By a NFPA-qualified person ☐
(b) For at least 30 minutes after the operation has finished ☐
(c) By at least 2 suitable persons together ☐
(d) By an API 1169-qualified person ☐

Chapter 16

The pipeline inspector's environmental responsibilities

16.1 Inspector environmental responsibilities

If you asked them, most pipeline inspectors (PIs) would probably consider themselves technical specialists rather than protectors of the environment. A background in welding, NDE or QA is normally not a good finishing school for environmental warriors. With this as a background, the API 1169 certification programme is unique among the 15 or so API ICPs in including a level of environmental and pollution control responsibilities into the inspector's role. This arises, no doubt, from the effect that overland pipeline construction projects can have on the environment, coupled with the fact that the PI's environmental responsibility is confirmed in RP 1169 itself so it cannot be ignored as a valid part of the API 1169 exam BoK.

Depth of knowledge?

Realistically, RP 1169 accepts that a PI is unlikely to be an expert on environmental matters. It therefore specifies the PI's role based on the assumptions that

- the PI needs a **basic knowledge** (only) of the environmental aspects of a pipeline construction project

but

- the PI does have clear **responsibility to verify** that those environmental procedures and protection that are in place are correctly implemented and monitored.

Expressed like this, you can see that it is not that much different to the monitoring and verification activities that the PI does on the technical and quality side of the project. The subject is a bit different but the activity is the same. There's further good news for the PI when you compare the nature of the verification of environmental protection procedures with those of the technical engineering world. The long and protracted discussions about the project environmental procedures will have all been played out and agreed at the project feasibility and planning stage, long before the PI becomes involved. The deep and wide nature of discussing anything to do with the environment means that, once a project procedure is eventually approved by all parties, then all details have been covered. There is nothing left to discuss, which makes implementing the procedure a fairly straightforward, deterministic affair. This is why a PI is able to verify and report on its implementation without too much trouble.

The inspector has another ally in environmental matters in all the statutory regulations that exist behind the scenes of a pipeline project. Environmental legislation is extensive and enthusiastically enforced by most jurisdictions, with their own structure of inspection, investigation and penalties that can be imposed. This acts wonderfully in concentrating the mind of the pipeline contractor in giving priority to environmental compliance matters. The result is that a PI can often find that there are fewer negative observations and non-compliances to be found in environmental activities than there are in the purely technical and quality parts.

The six main responsibilities

The six main environmental responsibilities of the PI are listed fairly logically in chapter 6 of RP 1169 as

- Section 6.2: Erosion, sediment and runoff control
- Section 6.3: Environmental permits
- Section 6.4: Major statutes (CFR, OSHA, CERCLA regulations etc.)
- Section 6.5: Water crossing permits
- Section 6.6: Use of natural water sources (for withdrawal and discharge)
- Section 6.7: Handling contamination issues.

There is nothing particularly to differentiate between these. Each refers to its own relevant statutory regulations but the job of the PI remains the same in each: *monitor, verify and report* compliance against the procedures put in place by the pipeline contractor.

Environmental protection: exam questions

The API 1169 BoK confirms that 15% of the 100 examination questions will be about environmental protection during pipeline construction. The breakdown of subjects within this 15% suggests that you can expect most (if not all) of these to be sourced from the regulations and other documents that make up the open-book section of the BoK. In the nature of regulations, these are heavily structured, written in formal language and distinctly unexciting to read. However, individual subjects are not difficult to find using the index and there is not too much interpretation left remaining in the statements they make.

There is no need to read through all these documents hoping to remember their chapter and verse requirements. You won't. Familiarity with the index of each is more important so you can find where to look for a particular exam question subject. We will look now at some of the referenced documents from the open-book part of the exam BoK.

16.2 FERC document: *Wetland and Waterbody Construction and Mitigation Procedures*

This is one of the two FERC (Federal Energy Regulations Commission) documents that specifically address environmental issues for pipeline construction programmes. Its stated purpose is to minimise disturbance due to pollution of wetlands and waterbodies. It is therefore essentially a mitigation procedure, restricting construction activities that can occur near waterbodies.

What is a waterbody?

A *waterbody* is defined as any river, natural drainage channel or lake containing water, either flowing or stagnant. They are subdivided into three sizes based on their largest dimension: ≤ 10 feet (minor waterbody), between 10 and 100 feet (intermediate waterbody) and > 100 feet (major waterbody). In contrast, a *wetland* is a bog, swamp or similar that is wet but has vegetation growing in it as well. The definition excludes any cultivated cropland. Waterbodies and wetlands contain living plants, fish and animals and may or may not be of particular use to society but they are nevertheless all protected in a similar way.

Contents of the document

This is a short 20-page document, fairly easy to read and search for specific information required. Important requirements it raises are as follows.

- **Authorisation** – formal approval is required from the FERC. Several specific permits have to be obtained before activities can start.
- **Planning** – particularly spill prevention to prevent water contamination by fuel or hazardous substances. Rules restrict what can be done within 100 feet of the water's edge.
- **The environmental inspector (EI)** – this is a specialist assigned role, separate to that of the PI.
- Specific requirements for **crossing procedures** covering trench excavations, work and spoil areas, sediment control etc.
- **Control of hydrostatic testing** – particularly extraction and discharge of water to minimise any problems of erosion, sediment or drainage.

There are some differences in the requirements for the three sizes of waterbody and wetlands, but the principles are much the same and easy to track down open-book in the document.

Major waterbody crossings

Crossings of major waterbodies (rivers etc. > 100 feet wide) carry various practical restrictions for a pipeline construction project. Figure 16.1 shows some of them. For this length of crossing, pipeline contractors will normally use the 'dry ditch' method. As the name implies, this involves laying the pipeline 'dry' not just excavating a trench across the waterbody and lowering the pipeline down through the water into it. Figure 16.1 shows three alternative ways of doing this. Individual restrictions are placed on each one, again with similar objectives of minimising disturbance and contamination of the water and land area immediately surrounding it (Photo 16.1).

Restoration of the crossing area after construction is finished is always an important FERC consideration. Waterbody banks have to be stabilised, erosion control barriers etc. installed and everything returned to as near its pre-construction state as possible. In practice, this activity may continue after the pipeline project is complete and the PI's involvement has expired.

The pipeline inspector's environmental responsibilities

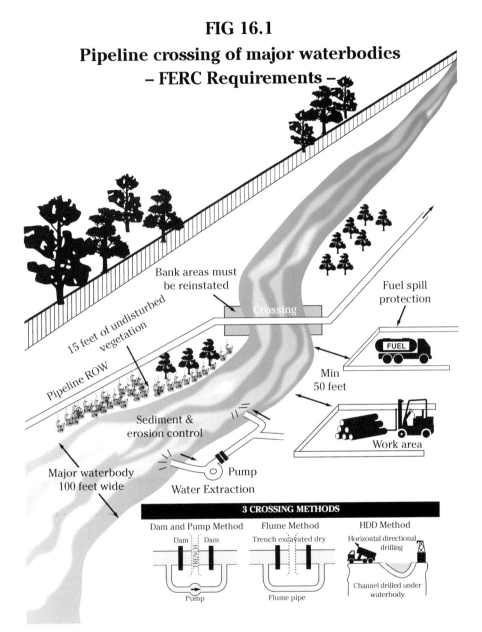

FIG 16.1
Pipeline crossing of major waterbodies
– FERC Requirements –

Photo 16.1 Pipeline excavation near waterbody (photo courtesy 123RF)

Wetland crossings

Wetlands can be much larger than waterbodies, perhaps extending for miles along a pipeline RoW. There are different types of wetland defined by NWI (National Wetlands Inventory) classification. The FERC mitigation procedures are not that different from those for waterbodies; the restoration plan can be more involved, however, requiring specialist environmental knowledge input.

Exam questions: what to expect

This is an easy document in which to find the answers to API 1169 open-book exam questions. It's the only document dealing specifically with wetlands and waterbodies so is easy to identify as the source for any exam question that mentions them, even if the question doesn't specifically mention FERC as a clue. The style of the document encourages straightforward factual questions about what it says, without much room for interpretation or fancy word-games. There are also only five or six different sections in the index, so things are easy to look up. The following subjects make convenient exam questions, so are worth looking up so you know where to find them.

- Which construction activities *authorisation* (i.e. permits) is needed for, and who to apply to.
- *Separation distances* from waterbodies/wetlands required for various work activities.
- *Notice periods* that have to be given for this and that.

16.3 FERC document: *Upland Erosion Control, Vegetation and Maintenance Plan*

This is the second of the FERC documents addressing environmental mitigation for pipeline projects. It addresses the situation where the pipeline RoW passes through *upland* regions of the landscape. There is no definition of 'upland' given in the document (i.e. height above sea level), so the inference is that it can apply to landscape at any height, with the provision that the land is inherently dry (i.e. not a waterbody or wetland). The content of the document seems oriented towards precautions to be taken on sloping ground, although it does not actually exclude flat landscapes such as plains and desert.

The main problem: erosion

The central problem that the document addresses is the problem of soil and land erosion that happens when a landscape is disturbed in some way. Pipeline RoWs traverse landscapes over long distances and, once vegetation is removed in preparation for excavating a trench, wind erosion quickly carries the soil away. For sloping landscapes, erosion by water is the biggest problem. Water-eroded soil is removed downhill and deposited as unwanted sediment somewhere else. The higher the altitude and slope, the greater the problem. In both cases the change to the landscape, once it has happened, will be permanent.

Document content

Similar to the wetland/waterbody FERC document, this one starts with the authorisation/planning required and moves on to mitigation actions required during the excavation, construction and land restoration stages of the project. It contains more procedural detail than for waterbodies/wetlands, mainly because there are many more activities (and length of pipeline) involved when the RoW passes over upland than water. Most onshore pipelines will be traversing land for 95% + of their length with a few widely spaced water crossings where they can't be avoided en route. Typical land-based features that have to be anticipated are

- existing upland drainage systems
- roads, transport and habitation
- vegetation and animals
- multiple types of subsoil and topsoil
- rain and snow.

Owing to the long overland length of the RoW, it follows that more requirements for wayleaves, permissions and permits will arise along the way. This increases the number of parties that have to be communicated with, and so more extensive communication and co-ordination procedures are needed. Other API 1169 BoK elements such as the Common Ground Alliance (CGA) best practice document 13.0 also address this.

Erosion control measures

Figure 16.2 summarises some of the erosion control measures imposed on upland pipework routes. During excavation, the RoW and surrounding slopes must be stabilised to minimise water runoff erosion. Natural or synthetic mulch may be used. Off-road vehicles are restricted to pre-agreed routes, with signs, cameras etc. to ensure these are correctly enforced. Once the trench is dry then temporary slope-breakers are installed to segment the trench into sections, preventing water runoff traversing the full length. On steeper slopes, permanent ones are left in place. Depending on the site, soil types excavated from the trench and surrounding site work areas are normally segregated for correct re-use.

Following laying of the pipeline and backfilling of the trench, the general requirement is for soil reinstatement to be completed within 20 days. The site is then revegetated by mulching and seeding of plants and trees. Timing is limited by seasonal considerations but the objective is to get it done as quickly as possible before erosion has time to occur. On steep slopes, permanent slope-breakers may need to remain in place, particularly if any regrading has taken place during the project.

Finally, the EI

Both the upland and waterbody FERC documents recognise, and indeed specify, the existence of a site EI with the responsibility of monitoring and reporting on these environmental mitigation parts of the project. A list of 17 duties is specified, stretching from authorisation and planning through to the final site restoration and monitoring

FIG 16.2
Upland erosion control
– Some FERC requirements –

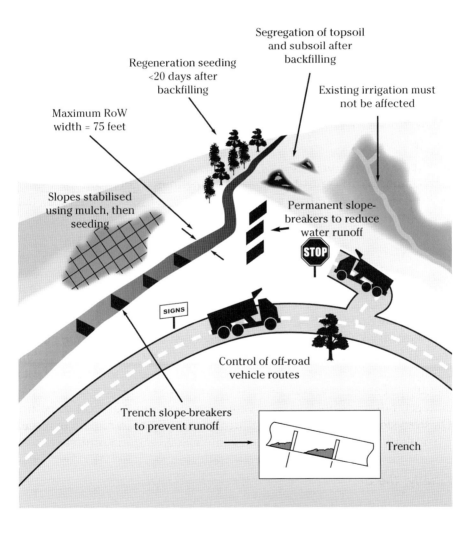

phases. All of these relate to environmental issues and they don't have the wider technical project quality remit of the RP 1169 PI. Although the EI role does not contradict the role of the PI it makes sense to assume that they are different people. The PI still has the overall responsibility to monitor environmental aspects but, in practice, will be able to rely on the specialist knowledge of the EI for day-to-day support.

Exam questions

These will follow the same guidelines as for those taken from the FERC wetland and waterbody document; see earlier.

16.4 The national pollution contingency plan: 40 CFR 300

A mainstay of environmental protection in the USA is document 40 CFR 300, its full title being the *National Oil and Hazardous Substances Pollution Contingency Plan*. As a federal regulation, it applies to all industries, not just pipelines, and is one of the most powerful statutes, placing firm requirements on facility owner/operators to comply.

The objective

The objective of 40 CFR 300 (sometimes called the Pollution Contingency Plan or PCP for short) is to deal with instances of hazardous pollution *once they have happened*, rather than prevent them happening in the first place. It sets out an organisation-like structure of national bodies and agencies to decide and implement remedial action, and generally administer the problem. As a Code of Federal Regulations document, 40 CFR 300 has national effect across the USA. Its existence supports the requirements of the Comprehensive Environmental Response, Compensation and Liability Act 1986 (CERCLA). Owing to the importance of the subject and the powerful organisations representing both regulatory bodies and polluters, the regulations are expansive and complex. 40 CFR 300 fits this bill well – it does not make light reading, being written in legal rather than technical language. To support this it needs long lists of acronyms for the national and regulatory bodies involved, all fitting into a nested administration structure of who is responsible for what.

What's this got to do with the RP 1169 PI?

Very little, most of the time. The PI is highly unlikely to have any real involvement with the management of the clean-up activities after a pollution event. The compulsory involvement of regulatory authorities pushes issues to a higher level, often out of the hands of owner/operator and construction site personnel. Public relations, media advisors, legal advisers and even shareholders' representatives appear out of the woodwork once a pollution incident occurs.

The 40 CFR 300 document itself, as a federal regulation, has no recognition of the RP 1169 PI or any other manufacturing or construction inspector; it mainly refers to bodies and organisations when defining what their responsibilities are. Overall, this is good news for the API 1169 examination, as it limits the use of 40 CFR 300 as a valid source of exam questions.

The actual role of the PI is set out in RP 1169 clause 6.4.5. This is only two lines long and limits the familiarity of the PI to the way in which 40 CFR 300/CERCLA impact on the owner/operator's obligations. This is as good as saying that the PI will not be questioned on the detailed content of the documents or the long lists of organisations and related regulations cited within them. The published API 1169 exam BoK (which confirms that 15% of the exam questions will be about pipeline construction environmental protection remember) contains only a single bullet point (of the 13) that mentions spills/incidents. This makes sense as the emphasis is on *preventing* incidents rather than cleaning up afterwards.

40 CFR 300: the parts in the BoK

Only two sections of 40 CFR 300 are included in the API 1169 exam BoK:

- Part A: *Terms and definitions* (24 pages of them)
- Part E: *Response* (to pollution incidents).

Part A, *Terms and definitions*, as is usual for API-style exams, contains vastly more information than you need to answer API 1169 exam questions. There are numerous regulatory bodies, with their agencies, and trustees and committees and the acronyms that represent them. It is difficult reading and you won't learn much. The reason why it is included in the BoK is probably for grammatical completeness (i.e. to define any terms that appear in the other Part E *Response* text, if you did need to look something up).

Part E *Response* seems to be the important part. It forms a chronological summary of the activities that are required following a release of pollutants into the environment, i.e.

- discovery of the incident
- site evaluation
- removal action
- remedial programme
- remedial investigation
- remedial action.

From a PI's viewpoint these activities all carry much the same effect, which is the imposition on the pipeline owner/operator of a raft of external bodies who govern how and when mitigation and remedial actions occur. The owner/operator is required to proactively co-operate at all stages, starting with reporting the pollutant release and then providing all necessary site information to the investigation as the steps proceed. It is the role of the PI to be aware of these requirements placed on the owner/operator and also to co-operate if requested, on their behalf.

In summary of how things work, a *lead agency* is placed in charge of the incident. *Site permits* will be required to do various things and *statutory reports* have to be made at ongoing stages. Some activities may attract CERCLA funding under the control of a *National Priorities List (NPL)*. Communications between all the parties are a vital part of the process. Concerns have to be addressed by written responses and there are regulated procedures for initiating, approving and disseminating these. In the centre of this storm sits the pipeline owner/operator, so you can see why their co-operation is so important.

What about 40 CFR 300 exam questions?

There are unlikely to be more than one or two in the exam due to the reasons discussed previously. They will be open-book questions with the questions phrased so that it is quite clear that a pollution incident has already occurred. This points you clearly towards RP 1169 (6.4.5) and its citing of 40 CFR 300. This document *is awkward* to pick meaningful questions from. We know that the PI is not mentioned in the document so the most likely tack is towards questions about the responsibilities of the owner/operator. Alternatively, a more contrived question on the meaning of one of the term definitions in Part A provides an easy refuge for any question-setter that has run out of ideas. There are lots of these definitions – it's worth looking up the following ones as they are the easiest to phrase an exam question around.

- Hazard Ranking System (HRS)
- Hazardous substances
- Lead agency
- Neighbouring
- On-site Coordinator (OSC)
- Release
- Removal
- Support Agency Coordinator (SAC)

Question set 16.1 at the end of this chapter contains a set of questions to help you to look through 40 CFR 300. Remember that only Parts A and E are included in the API 1169 BoK.

16.5 Protection of navigable waterways

As pipeline routes frequently pass or cross waterbodies there is special government protection put in place to protect their effect on those waterways that are either tidal or navigable (i.e. used for commerce by ships or barges). This has the dual function of protecting against adverse effects on both commerce and the environment. Two federal documents are included in the API 1169 exam BoK to cover the subject:

- 33 USC Chapter 9, explaining general protection measures
- 33 CFR 21, which is about permits for dams and dikes.

These are both short documents (a few pages each). They fit into the open-book section of the exam BoK and, taken together, make up a small component of the 15% of exam questions devoted to environmental protection issues. However, in the real world of pipeline RoW planning they play a greater role, imposing a system of approvals and permits on pipeline operators that wish to route their pipeline over or alongside navigable waterbodies.

33 USC Chapter 9 contains only a few sub-sections (401, 403, 407a and 409) that are included in the exam BoK. It sets the requirement for government approval and permits for construction of, or work on, bridges, causeways, dams and dikes that cross or affect navigable waterways. Here's what the terms mean.

- **Bridges** cross waterways at a height, using a structure of some sort. They stay dry at all states of the water flow.
- **Causeways** (strictly) are built up from ground level and may become flooded at high tide etc. It's not uncommon for structures that are actually bridges to be called causeways. The Florida Keys causeway is a good example.
- **Dams** are obstructions *across* a waterway to prevent flow and hold back water for some reason. They can be wet on both sides.
- **Dikes** run *alongside* a waterway, their purpose being to prevent flooding. They are therefore intended to be wet on one side and dry on the other.

The main thrust of the document is to prevent unpermitted obstructions to waterways by defining approval and permit requirements for work related to navigable waterways. This also includes changes to existing structures, dumping of excavation spoil etc. Approval involves high-level agreement of the US Secretary of Transport, the Chief Engineer's Office and the Secretary of the Army, depending on whether the proposed work is within one US state or qualifies as an interstate project.

33 CFR 321 *Permits for Dams and Dikes in Navigable Waters of the United States* is limited to dams and dikes only. Similar to 33 USC Chapter 9, it specifies the need for permits, again with various different approval routes if it is an interstate issue.

The role of the PI

As with all regulations of this type, there is no mention or recognition of the PI in either of the two documents. The reason for including them in the API 1169 BoK is mainly to make the PI aware of the extent of permits required. Although the exam BoK does not mention navigable water permits directly it is the role of the PI to verify that *all* necessary permits and approvals are in place during the construction project, so environmental permits are no exception. In reality, the PI will not normally have been involved in discussing and agreeing the scope of the permits (they are done well before, in the project planning stage), but will be in a position to ensure that they are complied with. As a presence on site, the PI is in a position to report on any activity that falls outside the permit scope. Typical real-world examples that occur in pipeline construction projects near waterbodies are as follows.

- Temporarily obstructing navigation channels by dredging or pipelaying equipment.
- Depositing refuse in navigable waters or on the banks (more common) where it can be washed away.
- Work starting before permits are officially approved.
- Misinterpreting any special conditions imposed by permits. These are often well understood and agreed at the planning stage but can be 'forgotten' by site contractors when the project is in progress.

Exam questions

You can expect a question or two in the API 1169 exam about the protection of navigable waterways. The two BoK documents are very small but the subject is an important one. As open-book questions you should find them easy. Have a try at question set 16.2 and see how you do.

16.6 Protection of birds and endangered species

Two short sets of regulations dealing with the protection of birds and endangered species of animals and plants are included in the API 1169 BoK. They are:

- *Endangered Species Act 1973* (sections 3, 4, 7, 9, 10, 12)
- 50 CFR 21 *Migratory Bird Permits* (subpart B).

As with much of the environmental protection context of the BoK the influence of these two sets of regulations is likely to be felt more at the pipeline RoW planning and clearing stages, than later on in the construction project.

The Endangered Species Act 1973

The main purpose of the sections (32 pages) included in the exam BoK is simply to prohibit interference with any animals or plants that are considered an endangered species. There is a lower category of 'threatened' species, but the principles are much the same. Which species are, or are not, endangered is the subject of much discussion and changes from year to year. There is a list available on the US Department of Fish and Wildlife website and a Federal Register document to go with it. Section 7 of the Act covers communication and interaction between the various government departments and committees changed with implementing the Act. Such high-level discussions are outside the remit of the PI and so have little relevance.

In a real-world pipeline project the involvement of the PI is likely to be limited to the following.

- Reporting any interference with animals or plants that may infringe the Endangered Species Act requirements.
- Checking that any permissions or permits that have been agreed are in place. This will generally have been done at the RoW planning stage but issues can still arise later during RoW clearing.
- Monitoring and verifying the taking of any scientific samples that may be requested by government agencies or environmental specialists.

As a PI it's not a good idea to get involved in discussions about whether a particular species is endangered or not. This is a job for the specialists. Out of interest, you may like to know that among the endangered species are the pink fairy armadillo, the butterfly-headed marmoset, the Queensland hairy-nosed wombat and the good old dibbler (a speckly little marsupial). Not so lucky are those species that are believed to be already extinct, including the broad-faced potoroo, the darling downs hopping mouse and the gloomy tub-nosed bat. No-one's really certain, however, as it takes 50 years of not being sighted before a species is officially classed as extinct. Recently reincarnated members of the animal kingdom include the hula painted pig, the Lord Howe Island tree lobster (or *Dryococelus australis*, to give it its proper name) and the Javan elephant (how exactly do you miss one of those?). Like I said, leave it to the specialists.

Protection of migratory birds

Friday 1 November 2013 was probably a day of upheavals at the offices of the US Fish and Wildlife Service, Department of the Interior (Federal Register Department). Unlike any other Friday, this one saw the official publication of the Federal Register Ed. 78, No. 212, containing the final version, *no less*, of the revised list of migratory birds. Essentially it lists more than 1000 bird species that are considered *migratory*, and therefore come under the remit of special legislation that is there to protect them. At the centre of the storm was, as usual, not huge discoveries of new migrating types clouding the skies but rather the subdivision and classification of species whose migration is already known about. This is the science of *taxonomy*. With the stroke of the taxonomer's pen the hoary redpoll, blue-winged warbler and boreal chickadee were added to the list of migratory species while the laughing gull and grey-headed chickadee were unceremoniously dumped off it, again for taxonomy reasons. No longer for those, the protection of Code of Federal Regulations 50 CFR 21. A bit less laughing and greyer feathers for them perhaps.

What is 50 CFR 21?

50 CFR 21 *Migratory Bird Permits* constitutes the official regulations that mean permits are needed before anyone can trap, kill, direct, track or otherwise interfere with any species of *migratory* bird. That's why the taxonomy is important. Being migratory is an essential part of the bird world. The banded arctic tern *Sterna paradisaea* (grey, black head, red beak) thinks nothing of its 50,000+ miles annual twice-round-the-world circumpolar journeys. Smaller birds do it too; the 'pocket-rocket' Rufous hummingbird travels 3000 miles between Mexico and Alaska to get the best flowers. That's a lot of wing flapping.

Once protected by 50 CFR 21, migratory species cannot be interfered with by pipeline construction projects, except under special permits that, you might anticipate, are not so easy to get. Pipeline RoWs through breeding or feeding grounds are generally not allowed. Only subpart B of 50 CFR 21 is included in the API 1169 BoK. It's short (seven pages) and covers

- general permit requirements
- exceptions for captive-reared ducks (mallards) – no-one seems to like these for some reason
- restrictions on how to remove birds from property or structures.

Exam questions

As with other bits of environmental legislation, these regulations on the protection of endangered species and migratory birds are notoriously awkward to source API 1169 exam questions from. Their format, content and tentative connection to the work of the PI simply makes it difficult. It's still part of the BoK, however. Question set 16.3 gives you a few practice questions to help you find your way through 50 CFR 21. Have a quick look through the Endangered Species Act 1973 and see if you can anticipate a few exam questions for yourself.

Question set 16.1: 40 CFR 300

Q1. 40 CFR 300: definitions

Which of the following would not be considered as a release of a pollutant in accordance with the definitions in 40 CFR 300?

(a) The release is accidental rather than deliberate ☐
(b) The release only affects persons solely in a workplace ☐
(c) The release only affects persons solely outside a workplace ☐
(d) The pollutant is abandoned in sealed barrels or similar containers ☐

Q2. 40 CFR 300: terminology

In national pollution contingency plan 40 CFR 300 terminology, areas of glacially formed wetlands located in the upper Midwest and usually occurring in depressions that lack permanent natural outlets are called

(a) Texas coastal Prairie wetlands ☐
(b) Prairie potholes ☐
(c) Pocosins ☐
(d) Western vernal pools ☐

Q3. 40 CFR 300: definitions

As mentioned in national pollution contingency plan 40 CFR 300, a data management system that lists and tracks releases addressed by the CERCLA superfund program is

(a) SEMS ☐
(b) SAC ☐
(c) SARA ☐
(d) SERC ☐

Q4. 40 CFR 300: CFR 300: part E: response

Persons in charge of a massive vessel or facility that discover a release of oil or hazardous pollutant must report the initial release directly to

(a) Environmental Protection Agency ☐
(b) United States Coast Guard ☐
(c) National response centre ☐
(d) CGA 'one-call centre' ☐

Q5. 40 CFR 300: part B: release remedial investigation

Which organisation has established the programme goals for feasibility studies and remedial investigation when a release has occurred and been listed on the National Priorities List?

(a) Department of Energy (DOE) ☐
(b) Department of Interior (DOI) ☐
(c) Federal Emergency Management Agency (FEMA) ☐
(d) Environmental Protection Agency (EPA) ☐

Q6. 40 CFR 300: part B: response

Which of the following would be considered as the correct cause of action following a release incident that is listed on the National Priorities List (NPL), in accordance with the remedial priorities of 40 CFR 300?

(a) Has been allocated a firm start date for remedial action ☐
(b) Has been assessed to be not immediately dangerous to public health ☐
(c) Has been considered eligible for fund-financed remedial action ☐
(d) Has been allocated as a federal-only responsibility rather than a state responsibility ☐

Q7. 40 CFR 300: part E: response

Under national pollution contingency plan 40 CFR 300 part E, who determines whether a release of oil or relevant hazardous pollutant constitutes a public health emergency?

(a) The lead agency ☐
(b) The Environmental Protection Agency ☐
(c) CERCLA ☐
(d) The National Institute for Occupational Safety and Health ☐

Q8. 40 CFR 300: umbrella legislation

An important piece of 'umbrella' legislation controlling many of the activities discussed in national pollution contingency plan 40 CFR 300 is

(a) USFWS ☐
(b) OSLTF ☐
(c) SEMS ☐
(d) CERCLA ☐

Q9. 40 CFR 300

The purpose of National Pollution Contingency Plan 40 CFR 300 is to provide an organisational structure and procedures for discharges of oil and

(a) Technical standards for repair or reinstatement ☐
(b) Environmental management of the consequences ☐
(c) Releases of hazardous substances ☐
(d) The setting up of a 'one-call centre' system ☐

Q10. 40 CFR 300: part B: response

The National Priorities List (NPL) is the list giving priority for

(a) Possible facilities at risk of having a release ☐
(b) Long-term releases that need assessment or remedial work ☐
(c) Data collection about historical releases ☐
(d) Organisation to act as 'lead agency' for response to releases ☐

Question set 16.2: Navigable waterways: 33 CFR 321/33 USC Chapter 9

Q1. 33 CFR 321: dams and dikes permits

Which of these is not classed as a dam or dike in navigable waters of the USA?

(a) A pipe bridge 20 feet high spanning a non-tidal waterway used for the passage of barges ☐
(b) A pipe bridge over a tidal waterway currently carrying no waterborne traffic ☐
(c) A weir ☐
(d) A pipe bridge over navigable waterway that is tidal ☐

Q2. Navigable water CFRs

The need for permits for creating dams and dikes in navigable waters of the USA is covered by

(a) 33 CFR 312 ☐
(b) 33 CFR 321 ☐
(c) 33 USC Chapter 9 ☐
(d) 40 CFR 300 ☐

Q3. 33 CFR 321: navigable waters dam/dike permits

In the USA the conditions to be imposed in any instrument of authorisation will be recommended by

(a) Local state governors ☐
(b) District engineer ☐
(c) Federal government (civil works department) ☐
(d) Assistant Secretary of the Army (civil works) ☐

Q4. 33 USC Chapter 9: obstruction of navigable waters

The creation or continuance of an obstruction, not previously affirmed by law, of navigable waters in the USA

(a) Is not a criminal offence ☐
(b) Shall be approved by the District Engineer ☐
(c) Requires the authorisation of the state governor ☐
(d) Is a criminal offence ☐

Q5. 33 CFR 321: navigable water dam/dike permits

In the USA, who decides if DA authorisation for a dam or dike involving an interstate navigable waterway will be issued?

(a) The Assistant Secretary of the Army (civil works) ☐
(b) The federal government (civil works department) ☐
(c) The level state governor ☐
(d) District engineers ☐

Question set 16.3: Migratory bird permits: 50 CFR 21

Q1. 50 CFR 21: migratory bird permit birds in buildings

Under the rules of 50 CFR 21, you must have a permit from your regional migratory bird permit office to move which of the following from a building?

(a) All migratory birds ☐
(b) Bald eagles ☐
(c) Mallard ducks ☐
(d) All birds ☐

Q2. 50 CFR 21: migratory bird permit

Bird species on the federal list of threatened or endangered wildlife need a special permit if you wish to remove them from premises. Details can be found in

(a) 50 CFR 17.21 and 17.31 ☐
(b) 33 USC Chapter 9 ☐
(c) 33 CFR 321 ☐
(d) 40 CFR 300 ☐

Q3. 50 CFR 21: migratory bird permit

Which of the following actions would you consider as the first thing to do if you find a migratory bird in a building essential to a construction project?

(a) Find out what species it is ☐
(b) Shut the door and call the local jurisdiction/authorities ☐
(c) Apply to the Department of the Interior (DOI) for a removal permit ☐
(d) Try to frighten it away ☐

Q4. 50 CFR 21: *migratory bird permit*

Which of the following statements regarding the activities of migratory birds should be considered as incorrect?

(a) Mallard ducks possessed in captivity without a permit shall be marked prior to six weeks of age ☐
(b) You must use humane methods to capture the bird or birds ☐
(c) You must immediately release a captured bird that has become injured back into the wild suitable for the species ☐
(d) You must seek the assistance of a rehabilitator prior to removing an active nest ☐

Q5. 50 CFR 21: *permit exception*

Under 50 CFR 21, migratory bird permits are generally not required for

(a) Any migratory bird classed as verminous or that is ill or infected ☐
(b) Any bird that exists in large flocks ☐
(c) Golden eagles ☐
(d) Captive-born mallard ducks ☐

Chapter 17

The pipeline inspector's quality responsibilities

17.1 Quality management: evolution or revolution?

The parallel universe that is quality management cannot, it seems, decide whether it is governed by evolution or revolution. Its principles have evolved over the past 40 years or so with standards like ISO 9000 hanging on through time, changing their format, content and even their titles. All this to please the additional countries, institutions and committees that wish to be associated with them, or not, as the case may be. In the fevered search for yet more things to comply with, standards of similar format have come along, offering compliance with the environmental, information security, personnel and even ethical requirements set out within their hallowed pages.

From time to time the relative peace of this evolution gets disturbed by some revolutionary new idea that comes along; the dull management equivalent, you could suppose, of the new kid on the block. Total quality management (TQM), Just-in-time (JIT), Six Sigma (6σ), X-stream leadership (XSL no doubt) and the Management by deciding nothing method (MBDN). Actually, I made that last one up, but you get the idea. These come and go, flower for a while, then become so successful that they are quietly replaced by something else. Slowly, the status quo returns, with a few more recognised acronyms to add to the excitement.

All of this fluttering away in the background does little to affect the less fluffy world of the mechanical plant or pipeline inspector (PI). The mainstream API ICP certificates API 510/570/653/SIFE/SIRE give most of their attention to the role of the inspector in technical issues such as materials, welds, corrosion, pressure and performance testing and the like, so all is well.

Unfortunately, for the second time so far in this chapter, that last bit isn't completely true either. The RP 1169 document and the exam BoK that goes with it take a completely different view of the relationship of the RP 1169 PI with the world of quality management/quality assurance and all the related terms that go with it. Please read on.

The PI as quality manager... surely not?

Possibly yes, according to RP 1169. Section 4.3 of the document says

- *Inspectors are expected to be the **principal** means of assuring work and material quality during field construction.*

This seems fairly clear. It doesn't give the PI the job of quality manager as such, but confirms the role of monitoring and verifying from a higher level that other quality managers are doing the job properly. Reading on in RP 1169 (section 4.3) confirms the wide scope of the items under the influence of the project quality management system, including standards, contract terms, specification, drawings – just about everything.

As always in API codes and RP documents there is a close link between the inspector and documentation requirements. Section 7.20 of RP 1169 sets this out. Once again, this reflects the PI's wide role in project planning, safety and environmental matters, not just technical ones. Remember that the PI's role is not to instruct or direct the contractor's QA or other staff. This is clearly built in to the wording of section 4.4 as

- *Inspector... must not direct nor supervise the contractor's work*

There may be a whisker of contradiction with the message of section 4.3 in this, but that's not too unusual in the world of QA double-speak. Either way, both statements make good API 1169 exam questions.

How many quality principles questions are there in the API 1169 exam?

Precisely 10%, says the official API 1169 BoK document. They are all closed-book questions and sourced (hopefully) from the various codes and documents that are included in the full BoK. Ten questions from the 100 in the exam paper are sufficient to recognise this as a not-to-be-ignored part of the BoK. It is feasible to expect, for example, that there will be more questions on quality principles than on individual technical subjects such as welding, NDE or coating inspection.

Exam questions on quality principles can be sourced from the two specific quality standards in the BoK:

- ISO 9000: *Quality Management Systems – Fundamentals and Vocabulary*
- API Q1: *Specification for Quality Management System Requirements for Manufacturing Organisations for the Petroleum and Natural Gas Industry.*

In addition, there is no doubt that a question or two can be scraped from the single paragraph in section 4.3 of RP 1169 and from loosely related topics such as documentation and records scattered around the other documents.

From an exam-setter's perspective, these two quality management documents are not easy to get sensible questions from. Neither is specific to pipeline construction projects, or in fact any construction project at all. There is no recognition of the role of any construction site inspector either, so they can't allocate inspection responsibilities, or describe tasks, or anything like that.

17.2 ISO 9000 and API Q1

The relevant content of ISO 9000

ISO 9000 is a document of concepts and principles. Strictly, the API 1169 BoK only includes knowledge of the terms and definitions in ISO 9000 but in practice this incorporates most of it. Expanding from earlier editions, ISO 9000 has become more conceptual than prescriptive. Like all other things conceptual, the idea itself attains greater credibility than its material intentions (i.e. any finished thing). The next feature is the fragmentation of all the terms and definitions into separate sections, spread over about 20 pages, each looking much the same as the last. This subdivision means that definitions may be found across different chapters containing principles of activities, processes, systems, customer characteristics (whatever they are) and so on. Quality management documents may or may not interest you – either way, they are disorientating and not easy to learn from. Try it and see.

API Q1: what's in there?

API Q1 is a bit like what ISO 9000 used to be. It is aimed at manufacturers (as the title says) and proposes quality-based principles to follow. These are set out broadly following the chronological steps of a manufacturing process (i.e. ordering, goods inwards, manufacturing, testing etc.) and are clearer and less conceptual than ISO 9000. Terms and definition are all listed in section 3: easy to find and read.

On balance, exam questions of a sensible nature are much easier to source from API Q1 than ISO 9000. The fewer, shorter sections are less disorientating and all the tables in the document are not needed anyway. It's still a generic document, however, and makes no mention or recognition of anything outside the world of product manufacture.

17.3 The way forward – BoK tracker questions

Given the somewhat intangible nature of quality management subjects, the API 1169 BoK helps out by listing 20 bullet points that summarise the exam knowledge requirements. It makes sense to predict that the question-setters are given this list to work to, rather than looking at random through ISO 9000 and API Q1 looking for points that they believe PI candidates should know.

Figure 17.1 shows this scope as a table. The 20 BoK bullet points appear in the left-hand column, followed by a simplified statement on what each is about. The far right-hand column shows the number of the BoK tracker question in question set 17.1. There's a mixture of straightforward and more difficult questions in there. A few are awkward simply because the BoK documents don't say much about the bullet-point subject, rather than any issue of clever interpretation.

Try the questions in question set 17.1 and see how you do (the answers are at the end of the book in Appendix 1). Whether you get them right or wrong, make sure to look up the answer in the referenced document location given in the answers at the end of this book. By doing this and learning from them you should be well on the way to being able to answer the rest (10% remember) of the quality-related questions in the API 1169 exam.

FIG 17.1
Guide to the 20 API 1169 BoK tracker questions in question set 17.1

API 1169 BoK topic	What it's really about	Awkward points for the PI	Where to find it	BoK tracker question number
Management of change	The keeping of lots of (fairly uninteresting) records of changes	Changes to drawings/specification are easy to monitor, but the system is often incomplete and lags behind the construction	API Q1 (5.11)	1
Verification of personnel qualifications	Checking everyone has the necessary qualifications, but only when they are required	Just because someone says they are qualified doesn't mean it's true	ISO 9000 (3.8.12)	2
Enforcing project requirements	What power does the PI really have?	API Q1 is more to do with products than construction projects. ISO 9000 doesn't mention the word enforcement (anywhere)	RP 1169 (4.6)	3
Inspector's roles and responsibilities	Whether the PI's role is about looking at QA procedures instead of looking at pipes	ISO 9000 and API Q1 do not mention the role of any inspector at all. RP 1169 does	RP 1169 (4.3)	4
Record management: legibility	Avoiding faded photocopies that can't be read (material test reports are notorious for this)	Understanding material test reports in other languages	ISO 9000 – as a general principle	5
Record management: traceability	Material traceability is the main issue	Traceability can easily be lost when long chains of sub-contracts are involved	ISO 9000 (3.6.13)	6
Record management: retrievability	Being able to find records when required	Some people will be helpful when you ask to see records and some won't	ISO 9000 (4.5)	7
Record management: retention	How long are records kept for?	It's difficult to know what will happen in the future	Back to API Q1 (5.11)	8
Record management: document revision status	Preventing superseded documents being used by mistake	PIs need access to contractors' systems to be able to monitor this	API Q1 (4.4.3)	9
NCR handling: control of non-conforming conditions	Non-conforming products or activities need to be correctly managed so they don't grow out of control, or get forgotten	Project procedures about how to treat non-conforming conditions always may leave room for discretion	API Q1 (5.10)	10

API 1169 BoK topic	What it's really about	Awkward points for the PI	Where to find it	BoK tracker question number
NCR handling: reporting	Who does the PI issue their reports to?	There may be a Chief Inspector between the PI and the owner/operator	API 1169 (4.7)	11
NCR handling: disposition	Disposition means the way in which non-conformances are arranged relative to other things	ISO 9000 hardly mentions this, and neither does RP 1169	API Q1 (4.5)	12
NCR handling: corrective/prevention actions	When things are found wrong they should be corrected	Easy in theory. Difficult in practice	API Q1 (8.4.2)	13
NCR handling: close-out	NCRs should not be left open	Sometimes NCRs are left outstanding in the hope they will go away	API Q1, ISO 9000, RP 1169 do not mention it	14
Root cause analysis (RCA): purpose	Finding why something went wrong and a non-conformance occurred	There may be more than one reason behind a non-conformity	API Q1 (6.4.2)	15
Root cause analysis (RCA): defining root cause	Stopping non-conformances happening again	Mistakes have a tendency to repeat themselves, unfortunately	ISO 9000 (2.3.4.3)	16
Calibration: equipment calibration status	Manufacturing and test equipment needs to be properly calibrated	Everyone likes to talk about this as it's an easy subject	API Q1 (5.8)	17
Calibration: methods	A procedure is required so it is done properly	There are many different calibration periods	API Q1 (5.8)	18
Material presentation and handling: quarantine, tagging and ID	Non-conforming weld consumables can be a problem on pipeline projects	ISO 9000 is more about management-speak than material control	API Q1 (5.7.6.1)	19
Material presentation and handling: standard requirements	Watch out for substandard ('rogue') materials	There's a lot of it about in some countries	API 578 (not in BoK)	20

Question set 17.1

Q1. Management of change

In the absence of any specific legal or customer requirements, how long (minimum) should MOC records be kept for on a pipeline construction project?

(a) 2 years ☐
(b) 5 years ☐
(c) Until the pipeline is handed over to the owner/operator ☐
(d) For the life of the pipeline ☐

Q2. Verification of qualifications

Regarding personnel qualification on a pipeline construction project, *verification* means that

(a) Everyone is qualified ☐
(b) Competence has been demonstrated ☐
(c) Specified requirements have been fulfilled ☐
(d) Requirements for a specific intended use or application have been fulfilled ☐

Q3. PI authority to enforce

PIs are empowered to stop any work activity on a pipeline construction project that may result in danger to persons, property, environment or

(a) Delays to the project schedule ☐
(b) Substandard work ☐
(c) Any activity that is not correctly documented ☐
(d) All of the above ☐

Q4. RP 1169 PI's role in QA

Pipeline inspectors are expected to

(a) Be the principal means of assuring work and material quality during field construction ☐
(b) Act as a support to the QA management system during field construction ☐
(c) Direct and supervise contractors work under the site QA system ☐
(d) Formally audit the field construction QA system to the requirements of ISO 9000 and/or API Q1 ☐

Q5. ISO 9000 documentation requirements

Under ISO 9000, all records must be

(a) Accessible at the field construction site ☐
(b) Legible ☐
(c) Under revision control ☐
(d) All of the above ☐

Q6. Traceability definition

According to ISO 9000 definitions, *traceability* is the ability to trace the history, application or

(a) Quality of a product or service ☐
(b) Ability to achieve the desired outcome of a service ☐
(c) Location of an object ☐
(d) Dependability of something ☐

Q7. Control of records

Now look here, neither ISO 9000 nor API Q1 have ever heard of the role of the RP 1169 pipeline inspector. Treating this as a general knowledge question then, which of these statements would you say does *not* appear in RP 1169?

(a) Inspectors should complete required documentation in a timely, concise and accurate manner ☐
(b) Inspectors should verify the control of documentation in a way that meets the quality objectives of the project ☐
(c) Inspectors are expected to complete all documents required by 49 CFR 195.266 or 49 CFR 192 ☐
(d) Inspectors should confirm and acknowledge the test plan for a pipeline pressure test ☐

Q8. Document retention

According to API Q1, documentary records on a pipeline construction project should be established, controlled and retained for a minimum of 5 years

(a) Unless a shorter period is acceptable to customer requirements ☐
(b) Or 20 years if the pipeline contains flammable gas ☐
(c) Including those originating from outsourced activities ☐
(d) Excluding those originating from outsourced activities ☐

Q9. Document revision status

According to API Q1, superseded/obsolete documents

(a) Shall be removed from the point of use ☐
(b) Shall be destroyed ☐
(c) Should be destroyed ☐
(d) Shall be marked with red stamps or markers for paperwork and given 'denied access' status on software systems ☐

Q10. Control of non-conforming conditions

According to API Q1, the procedure for addressing non-conforming conditions shall include provision for them being accepted under customer concession (a common-enough occurrence). What does RP 1169 say about the role of the PI during this concession procedure?

(a) The PI should approve the concession ☐
(b) The PI should be informed of the concession ☐
(c) The PI should document the concession ☐
(d) Nothing ☐

Q11. Quality reporting

When working under a site Quality System to ISO 9000, who should the PI first report any non-compliances with the quality system to?

(a) The contractor's quality manager and/or auditor ☐
(b) The person responsible for the non-compliance ☐
(c) The Chief Inspector ☐
(d) The site project manager ☐

Q12. Disposition of non-conformity records

Records of non-conformances on a pipeline construction project shall (must)

(a) Be presented to the PI for approval ☐
(b) Have a procedure saying what to do about them ☐
(c) Have monthly audits ☐
(d) Be in the form of a paper record ☐

Q13. Corrective/preventative actions

Under API Q1, actions used to correct non-conformances (i.e. corrective actions) shall be covered by a procedure identifying requirements for

(a) Reviewing customer complaints ☐
(b) Identifying the person responsible for the non-conformance ☐
(c) Classification into those that do/do not affect the project schedule ☐
(d) Approval by the PI ☐

Q14. NCR close-out

A practical way of closing-out a non-conformance report (NCR) is to

(a) Have it validated by a third party organisation ☐
(b) Change the specification requirements to invalidate it ☐
(c) Obtain a customer concession ☐
(d) Pretend it never existed so everyone eventually forgets about it ☐

Q15 Root cause analysis

According to API Q1, identifying the root cause of a non-conformance forms part of the activity of

(a) Verification ☐
(b) Corrective action ☐
(c) Validation ☐
(d) Planning ☐

Q16. Defining root causes of non-conformances

Defining the root cause of a non-conformance in a construction project quality management systems (QMS) is the responsibility of

(a) The pipeline inspector ☐
(b) The QMS auditor ☐
(c) The owner of the QMS ☐
(d) The owner/operator's QA manager ☐

Q17. Equipment calibration status

When measuring or test equipment is found to be out of calibration, then

(a) All measurements taken since the last calibration should be repeated ☐
(b) The customer shall always be informed ☐
(c) It must be calibrated immediately ☐
(d) The validity of previous measurements should be assessed ☐

Q18. Equipment calibration methods

Regarding the calibration of testing, measuring and monitoring equipment, this equipment shall

(a) Be kept in 'quarantine' storage lockers ☐
(b) Be calibrated at least annually ☐
(c) Have the calibration status identifiable by the user ☐
(d) Be made so it is non-adjustable in use ☐

Q19. Material handling, tagging and identification

According to API Q1 for materials and products, the term *preservation* shall include/apply to material handling, identification, protection, packaging and

(a) Transportation ☐
(b) Inspection and testing ☐
(c) Environmental aspects ☐
(d) ROW reinstatement ☐

Q20. Material standard requirements

Metallic material, including weld consumables, can be checked to see if it has the correct chemical constituents using

(a) DAC ☐
(b) PMI ☐
(c) VI ☐
(d) ECT ☐

Chapter 18

Snippets

18.1 Specialist pipeline terminology

The pipeline project industry is noted for its unusual terminology and acronyms. If you look through all the documents in the API 1169 BoK you can find reference to more than 30 technical, industry and regulatory bodies and organisations, each with its own acronym. Organisations apart, there are also plenty of technical terms used specifically in the pipeline industry; most of the BoK documents have their own terms and definitions section near the front of the document or a glossary of terms at the back. For this reason they make valid exam questions; it's a good way of checking whether exam candidates have some previous experience in the industry.

Some of the more unusual ones are listed below. Have a look through and tick off those that you really know the meaning of. You can check the answers on the following page.

Abandonment
Anomaly
Backfilling
Barrel
Bell hole
Bioremediation agent
Breakout tank
Buckle
Centering
Class locations 1,2,3,4
Cleaning pig
Close interval surveys (CIS)
Code of Federal Regulations (CFR)
Common Ground Alliance (CGA)

Corrective Action Order
Critical habitat
Easement
Encroachment
Event tree
FERC
FONSI
Granny rag
Holiday
Launcher
Magnetic flux leakage (MFL)
Mulch
Notice of Probable Violation
Pigging
Pocosins
Right of way (RoW)
Rip-rap
Significant nexus
Sorbents
Spike test
Tailgating
Trenchbreaker
Utility pig
Wales
Western vernal pool

Here's the answers

Abandonment. The process and actions taken by a Company at the end of the useful life of a pipeline or pipeline facility to gain approval from the regulator.

Anomaly. Some feature of a pipeline section or weld etc. that normally should not be there. Many pipeline anomalies result during the pipe manufacturing process and don't affect the performance of the pipeline or its ability to function in a safe manner.

Backfilling. Filling dirt back into a ditch or hole that has been previously dug. The process of filling the trench where a newly constructed or recently unearthed pipeline is installed.

Barrel. A standard measure of a volume of oil, equal to 42 gallons.

Bell hole. A hole dug into the ground over or alongside a pipeline to allow the line to be examined and to provide room for workers to perform maintenance on the pipeline. It has an upside-down bell shape, wide at the top and narrowing to a smaller diameter around the pipeline to be examined.

Bioremediation agent. Additive deliberately introduced into an oil discharge to increase its rate of biodegradation.

Breakout tank. A tank used to temporarily store oil in a pipeline system.

Buckle. An anomaly that represents partial collapse of the pipe wall; usually caused by excessive bending or curvature being applied to the pipe.

Centering. A method of determining the approximate location of a pipeline leak. It can be done manually using gas leak detection equipment.

Class locations 1,2,3,4. A regulatory designation for natural gas transmission pipelines that indicates the level of human population within a certain distance on either side of the line. The class location of a pipeline is a factor in determining the maximum allowable operating pressure of the pipeline. Class 4 indicates the most heavily populated of the class locations, representing an area where buildings with four or more storeys above ground are prevalent.

Cleaning pig. A device placed inside a pipeline to remove unwanted debris from the inside of the pipeline (Photo 18.1). The pig can be drawn or pushed through a pipeline or moved through the line as a result of the flow of the product in the line.

Close interval surveys (CIS). A method of testing corrosion protection systems on pipelines. It involves inspection and electrical testing of the corrosion protection system every few feet along the pipeline to confirm the status of the protection system and to help identify mechanical damage to the pipeline.

Code of Federal Regulations (CFR). The CFR is the official compilation of the US federal regulations of general applicability and legal effect. The CFR is divided into 50 titles that represent broad topical areas that are subject to federal regulation. Quite a few of them affect pipeline projects.

Common Ground Alliance (CGA). A non-profit organisation dedicated to shared responsibility in ensuring public safety, environmental protection and the integrity of services by promoting effective damage-prevention practices.

Corrective Action Order. An order issued by the Pipeline and Hazardous Materials Safety Administration (PHMSA) if it determines that a particular pipeline represents a serious hazard to life, property or the environment. Such orders usually address urgent situations arising out of an accident, spill or other significant, immediate or imminent safety or environmental concern.

Critical habitat. Habitat inhabited by an endangered species of plant or animal.

Easement. An acquired privilege or right, such as a right of way, afforded a person or company to make limited use of another person's or company's real property. Oil and gas pipeline companies acquire easements from property owners to establish rights of way for construction and operation of their pipelines.

Encroachment. The unauthorised use of a right of way in violation of the terms by which the right of way was established (e.g. easement).

Event tree. A type of logic diagram/analysis tool that is used to identify individual events and event sequences that can lead to accidents. On the diagram, each individual event is shown on a branch connected to a trunk that leads to the accident.

FERC. The Federal Energy Regulatory Commission. An independent regulatory agency within the Department of Energy that regulates the transmission and sale of utilities such as oil.

FONSI. Finding of No Significant Impact. A document prepared by a federal agency, based on the results of an environmental assessment, showing why a proposed action would not have a significant impact on the environment and thus would not require preparation of an environmental impact statement.

Granny rag. Type of coating or method of coating a pipeline in the field rather than a factory-applied coating.

Holiday. A discontinuity or break in the anti-corrosion coating on pipe or tubing that leaves the bare metal exposed to corrosive processes.

Launcher. A pipeline component (pressure vessel) used for inserting an inline inspection tool, cleaning pig or other device into a pressurised pipeline.

Magnetic flux leakage (MFL). An inline inspection technology in which a magnetic field is induced along a pipe wall through the use of a smart pig. As the smart pig travels through the pipeline, measurements are taken of the magnetic flux density at the internal surface of the wall and changes in measured flux density indicate the presence of potential defects.

Mulch. A layer of material applied to the surface of soil to conserve moisture, improve fertility and health of the soil or reduce weed growth etc.

Notice of Probable Violation. An enforcement tool involving a notice issued by PHMSA.

Pigging. Using devices known as 'pigs' to perform various maintenance operations in a pipeline without stopping the flow of the product.

Pocosins. Evergreen shrub and tree dominated wetlands.

Right of way (RoW). The defined strip of land on which an operator has the rights to construct, operate and/or maintain a pipeline.

Rip-rap. Rock, wood matting or some other material used to reinforce riverbanks or any other waterway against water erosion.

Significant nexus. A waterbody that by virtue of its location affects some other waterbody in the region.

Sorbents. Inert insoluble material used to absorb oil from water using adsorption.

Spike test. A short-term, higher pressure hydrostatic test applied to a pipeline to find smaller-than-usual defects.

Tailgating. The act of driving on a road too close to the vehicle in front, such that the distance between the two vehicles does not guarantee that stopping to avoid collision is possible.

Trenchbreaker. A barrier placed across an excavated trench to slow the flow of subsurface water along the trench down a gradient (see Photo 18.2).

Utility pig. A mechanical pig sent through a pipeline that performs simple mechanical functions, such as cleaning the pipeline.

Photo 18.1 Cleaning pig (photo courtesy 123RF)

Wales. A type of vertical shoring sheet that prevents collapse of an excavated trench.

Western vernal pool. Seasonal wetlands located in California.

18.2 Pipeline mapping: the NPMS

The USA is fortunate in having a National Pipeline Mapping System (NPMS) showing the location of all gas or hazardous liquid pipelines across the country. This system comes under the jurisdictions of the PHMSA, part of the US Department of Transportation (DoT).

How can I access the maps?

The maps are all available online at www.npms.phmsa.dot.gov. Free access at a scale of 1:24000 is available to the general public via the public map viewer, with more detailed 'members only' access for pipeline operators and US government officials. Pipelines and their related facilities are colour-coded and also show the source of recent incidents (collecting this information is one of the roles of the PHMSA).

Photo 18.2 Trenchbreakers (photo courtesy Bigstock)

The purpose of the maps

The maps have a multitude of purposes, including infrastructure planning, emergency response planning, security and environmental controls. They are also useful as a tool for initial planning of new pipeline RoWs, showing potential crossing areas. Accuracy is now very high owing to GPS co-ordinates while older pre-GPS routes claim an accuracy of about $+/-500$ feet. The PHMSA is always careful to point out that these maps should be used for guidance only and are not to be used as an alternative to the 'Call 811 before you dig' system (see the Common Ground Alliance best practice document 13.0).

Figure 18.1 shows a typical map from the NPMS 'public viewer'.

18.3 The longest pipeline in the world?

There will always be some dispute as to which is the world's longest pipeline. Because pipe diameter can vary as pipelines split off into branches and sub-pipelines, there is no definitive way to measure them as a total single length. The claim for the longest crude oil pipeline has been made for the Interprovincial Line installation stretching across Canada from Alberta to Montreal, via Chicago, a distance of 2353 miles

FIG 18.1
A typical map from the National Pipeline Mapping System (NPMS) 'Public Viewer'

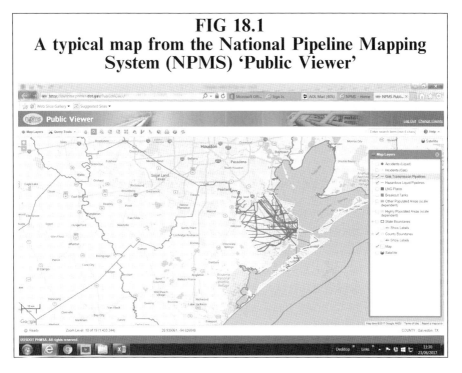

(3787 km). Similar claims are also made for the Druzhba or 'friendship' pipeline from Russia to the Baltic republics and Germany. This is more of a network than a single span, totalling 2500 miles (4000 km); its diameter varies from 17 to 40 inches (420 to 1020 mm).

China claims a 5410 mile (8656 km) national route from Xinjiang to Shanghai and Brazil a 3100 mile (4960 km) one from Mata Grusso do Sol (*Bushes of the South*) to Maranhao. Overall, there are more than three million miles of pipeline installed around the world, increasing rapidly year-on-year as oil and gas transfer increases to match supply with demand.

The immense length of long pipelines brings with it a mass of logistics problems. It is not the complexity or actual volume of the construction that's the problem, it's the fact that it is spread over a long physical route. Unlike construction work on a refinery site, which you could consider of comparable construction volume, the assembly does not all take place in a single location. On a construction site, road access has to be created from nothing, and then the landscape reinstated afterwards.

The main logistics problem is the number of pipeline lengths that need to be transported. The longest manufactured length of 42 inch diameter ERW (electrical resistance welded) line pipe is normally 80 feet, requiring 66 lengths per mile of RoW covered. Seamless pipelines are shorter (40–48 feet in length), weighing from about 13600 lb (6.2 tonnes) to 26400 lb (12 tonnes) per length depending on the wall schedule thickness, so you can't get large numbers of them on a road-legal truck. You can see why stringing the pipeline section out along the RoW is such a major logistic and expensive exercise.

18.4 Piping uphill, pipelines downhill

Piping welding versus pipeline welding

For the most part, the site welding of *refinery piping* and *overland pipelines* can be considered very different techniques. The difference starts with the codes involved. Refinery pipework is welded to the ASME B31.3 code using welders qualified to ASME IX. For overland pipelines, API RP 1104 covers both the welding technique and the qualification of the welders. Following this, some technical differences between the two situations start to take effect.

The main differences are as follows.

Refinery piping
- Mainly small diameter
- Variable thicker wall sizes
- Complicated runs with valves, fixtures and fittings
- Large inconsistent root gaps with hand-ground bevels
- Pipe inside is often inaccessible

Overground pipelines
- Large diameter (up to 60 inches)
- Relatively thin wall (many $<\frac{1}{2}$ inch)
- Long straight runs
- Machine bevelled with small ($\frac{1}{16}$ inch) root gaps
- Easy access to the pipe inside

These differences are instrumental in defining the type of welding used and the resulting speed of welding that is possible. Large-diameter pipelines, with their tight consistent root gap, mean that a downhand (also termed 'downhill') welding technique can provide the necessary weld penetration of the root faces, giving a strong joint. In addition, accurate alignment of the joint using an internal alignment clamp means that it is not necessary to add tack welds beforehand. Conversely, thicker wall ASME B31 piping with its hand-ground bevel has a larger inconsistent root gap. Complex shapes with bends and fittings also prevent the use of external alignment clamps, so tack welds are needed to hold the joint in alignment for welding.

Uphill versus downhill

If you can get a consistent root weld without the use of tacks then single-bead downhill welding is the quickest technique to use. The subsequent hot pass, fill passes and cap weld can be made in the same way, using simple cellulosic weld consumables. With a larger root gap and no alignment clamp, then tack welds are required. With this arrangement, a single-bead downhill root run would be inadequate to properly fill the root while providing reliable penetration into the root faces. To provide this, an uphill weaving technique is required to properly fuse the weld joint. This slower technique results in high heat input, requiring low-hydrogen weld consumables to avoid getting defects. This is the reason for the difference in technique between pipework and pipelines.

Summary

Refinery pipework is generally welded using an *uphill, weave* welding technique. Tacks, root pass and fill are usually done by the same welder. Overland pipelines, when they are welded manually, use a *downhill single-bead* technique, with root pass, hot pass, fill passes and cap weld being done in a fast production-line method using multiple welders who move progressively along the pipeline. Figure 18.2 shows the details.

18.5 Grizzly bears

Understandably, human interaction with grizzly bears (*Ursus arctos horribilis*) tends to be a rather one-way affair. At up to 10 feet tall and 1500 lb in weight they remain uncommunicative and a bit short-tempered in response to ecological impact surveys as to whether they mind having pipelines across their habitat. Of studies that have been done, notably in the Yukon and northwest Montana, results have revealed that they like to live in the bottom of valleys and other locations that are also suitable for pipeline routes (*Review of Oil and Gas Exploitation Impacts on Grizzly Bears* by Allen Schallenberger, University of Montana).

People who survey and map out the pipeline RoW are those who lose the most sleep due to grizzly bears. Left alone, bears are happy to spend their days doing bear-like things – the problem comes when they get disturbed. Bears like each day to be the same as the day before, with no strange visual or noise disturbances in their lives. Pipeline RoW surveys, with the presence of people and their noisy machinery, cause them

FIG 18.2
Piping welding versus pipeline welding

Refinery piping: ASME B31

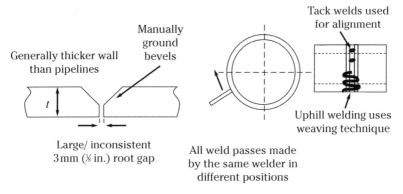

- Cellulosic coated electrode for root pass, low-H_2 rods for all the other passes

Overland pipelines: API 1104

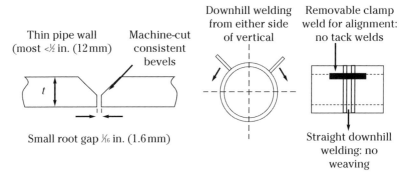

- Root and hot passes made before removal of alignment clamp
- All electrodes similar with cellulosic coating
- Different welders used for root, hot pass, fill and cap welds

confusion and distress. The worst thing of all is helicopters. They really don't like those.

Anyway, where this is leading is that it is the surveying then construction phase of pipeline projects that causes the most problems for grizzly bear tempers. Once the pipeline is completed, then they probably don't mind too much. They will actively avoid new areas of human habitat as long as they are not attracted by the availability of food waste or people actively feeding them. By the way, it's just not true that grizzlies like to walk along pipelines to keep their feet warm. Of those photographs that have been taken of bears walking along pipelines, they are just using it as a convenient way to cross a river or some other obstruction. That's cleared that one up.

Secondly, in support of the role of the PI, it seems to be a much safer job than that of a pipeline RoW surveyor in the habit of stepping into the forest off a helicopter to survey and map the pipeline route.

Thirdly, although grizzlies and other types of brown bear are officially not an endangered species, six of the seven or eight other types of bear are, so they fall within the protection of the 1973 Endangered Species Act that forms part of the API 1169 exam BoK.

Fourthly, there probably won't be any API 1169 exam questions specifically about grizzly bears. It's an interesting subject though.

18.6 Pigs

What are pigs?

Pigs are cylindrical mobile tools that move through a pipeline for various purposes including cleaning, scraping, checking the pipe diameter profile and inspection. They range in size from about 4 inches up to 60 inches in diameter for the largest pipeline sizes.

Why are they called pigs?

No-one is really sure, but two possible explanations are as follows.

- Early models were covered in leather and, when passing through a pipeline, someone decided they sounded like a squealing pig.
- Pig stands for *pipeline inspection gadget*. Sounds better, but probably equally unlikely.

FIG 18.3
Pigs

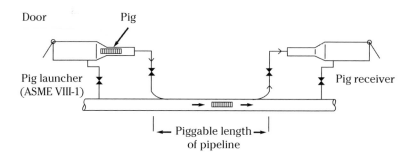

Types of pig:
- Mandrel pig with scraping brush
- Swab pig for diameter sizing
- Foam pig for cleaning
- Gauging pig with aluminium plates for ovality checks
- Venting pigs: pressurised by water to eliminate air before hydrotest
- Batching pigs: used to separate products in the pipeline
- Gyroscopic pig: provides a 3D isometric of the pipeline route using GPS
- MFL pig: detects defects using magnetic flux leakage
- UT pig: uses ultrasonic testing to accurately size defect

How does pigging work?

Piggable pipelines are fitted with a launcher and receiver vessel at either end of the pipeline. These are high-pressure coded vessels (e.g. ASME VIII) capable of taking full pipeline pressure and have a hinged door to enable the pig to be placed in or taken out. Figure 18.3 shows the arrangement. To be piggable, the pipeline itself needs to have bends of a certain minimum radius and no valve parts or other components intruding into the line to foul the pig. After being launched from the launcher vessel, the pig travels under the pressure of the fluid in the pipeline, typically at about 5 mph, over the full length of the pipeline until it is diverted into the pig receiver vessel at the other end.

Types of pigs

The simplest ones are sizing pigs, which simply check the diameter and roundness of the pipeline, and scraping/cleaning pigs, which clear the inside surface of deposits, hydrotest water and so on before the pipeline is recommissioned. More advanced 'smart' pigs use magnetic flux leakage (MFL) or ultrasonic testing (UT) techniques to detect and size wall defects in the pipeline. A tracking system logs and displays the size and location of defects found in a computerised display.

How much does pigging cost?

It's expensive. Smart pigging runs are very expensive as multiple pig types/runs are required.

H&S aspects

There are significant H&S aspects to pigging, so safeguards are required for

- interlocking of the launcher/catcher vessel doors to prevent them being opened when under pressure
- personnel exposure to pigging resides such as H_2S, benzene, toluene etc.
- venting/release of explosive hydrocarbons.

18.7 Mallard ducks

If you or I were a captive-bred mallard duck and a pipeline RoW was planned across our territory, we would have a bit of a problem in deciding the safest place to live. In addition, as we saw in Chapter 16, the 1973 Endangered Species Act doesn't seem too concerned with our welfare. Mallard ducks are the most abundant waterfowl in the world, laying 5–15 eggs at a time, being resistant to heat and cold and eating almost anything. Also called the wild duck, mallards are the ancestors to most other kinds of duck and can live in individual pairs or in flocks, called *sords* would you believe.

Incidentally (and this is relevant) only the female mallard, which is dull and brown with a yellow beak, goes 'quack'. More specifically, it announces its presence not by a single quack, but by a series of 2–10 quacks, starting off loud and getting softer and shorter – called a *descrendo call*. The male (shiny green with a yellow beak) doesn't quack, but makes more of a rasping noise, which is good to know.

Because there are tens of millions of mallard ducks in the world, breeding and hunting them is allowed, and doesn't seem to make much difference to the overall numbers. The 1973 Endangered Species Act (in the API 1169 BoK remember) doesn't get too involved with their protection. It does tell you, however, how they can be marked – by tattooing or chopping one of their toes off (yes it's in there in the text).

Anyway, getting back to the problem of where to live near the pipeline RoW. Ducks can roost more than a mile away from the waterbody where they feed, mainly at night. That's not much good, as construction work, lights and heavy machinery obstruct their daily patterns, as does the attention of all the potential new duck-hunters in the area. You can see the problem.

The word on the street (waterbody) is that wise ducks head for one of the couple of hotels in the world that are particularly duck-friendly (yes there are such things). There is one where, as guests of honour in the hotel (gratis of course), they live in a rooftop enclosure crafted from sculptured glass and marble and get fed on the finest tidbits. Every day, they are roused gently from their slumber by their own butler and then perform a formal march on a red carpet around the hotel lobby fountain. In another, a family of mallards lives in an enclosure in the hotel gardens and with freedom to roam into the foyer where they have a mini-fountain and automatic duck-feeder. By far the best, however, is one hotel where the hotel itself and room names are all duck-related. A group of ducks live like royalty in their own annexed duck house where, again, they have their own waiters. Even the restaurant is duck-themed and, before you ask, it is absolutely proud of its duck-free menu.

That's the place to be.

18.8 Earthquakes versus pipelines

The problem

Something like 12000–14000 earthquakes of magnitude 4 and above happen in the world every year along natural fault lines, 700+ of these being in the USA and Alaska. At the larger magnitudes, up to 8, the ground can move up to 20 feet horizontally and 5 feet vertically, along the route of the fault line. It is this offset of the ground that causes the main problem with pipelines. Above-ground lines slide, sway, rock and break while underground lines suffer crushing under the weight of the soil in which they are buried.

Pipeline failure mode

These are divided into the following three main factors after an earthquake has affected the RoW.

- **Leak tightness** – has the line fractured or distorted so much that it leaks at welds or flanges?
- **Operability** – is the line still capable of transporting process fluid?
- **Position retention** – is the line still supported on its supports, or has it slipped and fallen off?

Each of these three can be built into the pipeline project specification under the general heading of *seismic designs*.

General principles

The principles of seismic design are easy for pipelines as, unlike buildings, they have no significant height to 'sway', thereby increasing the forces. Under the influence of ground movement, a structure that resists bending (stiff) is more likely to fracture than one that is more flexible. Similarly, if the amount of vertical or lateral (sideways) movement can be reduced in some way, then forces and distortion will be reduced, improving the situation. For buried pipelines, the main problem is the crushing effect of the soil surrounding the pipeline so, again, if these forces can be reduced, the chance of failure is lowered.

Solutions

Figure 18.4 shows some solutions regularly used for pipeline design. Figure 18.4(b) illustrates a buried pipeline surrounded by expanded foam ('geofoam') blocks. These have one-hundredth of the density of soil so when the ground moves they crush easily, eliminating the external crushing pressure on the pipe.

Overground pipelines are shown in Figures 18.4(c) and 18.4(d). A zigzag pipeline route increases lateral flexibility and spring-loaded 'sway-braces' can be used to restrict the amount of sideways movement. Similar arrangements prevent excessive movement in the longitudinal direction. When lines cross major faults (the Denali fault on the trans-Alaska pipeline is a good example), the pipe can be mounted on a series of transverse slider bars (see Figure 18.4(d) and Photo 18.3). These allow a lot of movement without the pipe slipping off the supports.

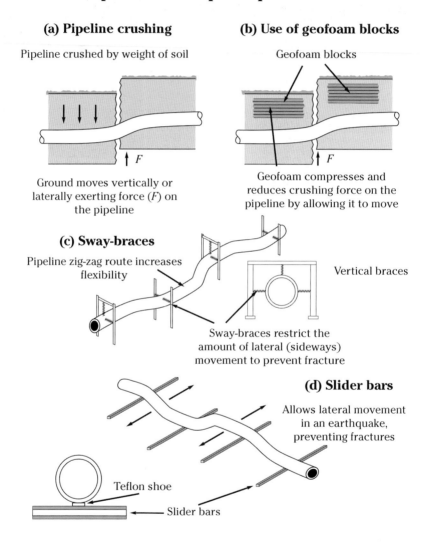

FIG 18.4
Pipeline earthquake protection

Photo 18.3 Trans-Alaska pipeline (photo courtesy Bigstock)

Codes and standards

Earthquakes are not a new problem so there are codes and standards around offering specifications to work to. Common ones are

- American Lifelines Alliance (ALA) *Guidelines for Design of Buried Steel Pipes*
- American Society of Civil Engineers (ASCE) *Guidelines for the Seismic Design of Oil and Gas Pipeline Systems*.

18.9 The pipeline environmental impact assessment (EIA)

In the majority of countries of the world any overland pipeline project must have an environmental impact assessment, or EIA, carried out at the planning stage, long before construction can be even considered. Almost everyone agrees that this is a good idea.

What is the purpose of the EIA?

Formally, it's to

- set out the biophysical and socio-economic impacts of the pipeline construction and operation and identify/propose any necessary mitigation measures.

Loosely translated, this means the effect that the pipeline will have on the animals, plants and people that will be affected by it. There is no limiting distance assumed, so it applies equally to those that are nearby or far away from the proposed pipeline route.

Who does the EIA?

Normally the pipeline operating company that wants to have the pipeline built and then operate it for their own commercial gain. However, the fact that it is an infrastructure project means that all manner of government departments and agencies, landowners and specialist consultants need to get involved. Regional and national governments ultimately have to approve the EIA before a project can become a reality. Such activities are rarely quick: 5–20 years is not unusual for large trans-national pipelines, with several EIA reviews or revalidations along the way.

What's in it?

The breakdown of subjects is fairly standard content. Although the remit of an EIA is to cover the impact of both construction and operational phases of the project, the construction phase normally takes most priority. The greater economic and societal effects of making large amounts of oil or gas previously available at point A now available at point B is left to government political advisers, economists, financial analysts and other such foretellers of the future. The more tangible impacts during the operational phase are included, as far as they can be anticipated and quantified.

Site erosion control
Land erodes once vegetation is cleared, and the erosion of banks, waterbodies and sloping trenches is an undesirable side-effect of pipeline construction. Mitigation measures are available for most of them.

Airborne emissions
Pipeline construction involves the emission of CO, CO_2, NO_x and CH_4 from project and road vehicles. During pipeline operation, leaks from valves (called fugitive emissions) have to be predicted. Construction also involves dust from earthworks, land clearance and materials handling.

Noise and vibration
Noise and vibration are mainly from construction activities such as clearing, ditch digging, drilling, blasting, pipe handling and vehicle

movements. Vibration can affect existing buildings and historical artefacts.

Effect on soils
The issues here are changes in soil structure and degradation of soil quality as a result of compaction or erosion.

Ecological impact
Vegetation, plants, birds and animals will inhabit both the 20–25 m RoW corridor width and areas remote from the working width. Endangered species must be identified and any effect on them mitigated.

Effect or groundwater
Groundwater sources can be affected by waste/chemicals, mainly during the construction phase.

Effect on surface waterbodies
This may be due to pollution during construction. The EIA normally specifies whether HDD or open-cut trenching will be used to cross waterbodies to keep impacts such as sediment transfer to a minimum. Permanent dams and dikes will have a continual impact throughout the life of the pipeline.

Traffic
Pipeline construction requires a large-scale transport operation to deliver pipework sections to the site. Heavy machinery is also required throughout the project phases of excavation, lowering-in and back-filling. Temporary access roads are essential features of most sites.

Waste
This is a construction issue, mainly to do with construction waste, rubble, scrap materials etc. Septic waste and wastewater generated from construction camps also needs to be managed.

Archaeology/cultural heritage
This gets subdivided into heritage items of archaeological, historical, religious cultural or aesthetic value. Sites of archaeological interest not discovered until after excavation has started can cause long project delays.

Population land use
The main issues are

- resettlement of people or communities from their homes
- loss of productive agricultural land.

The problem: impact rankings

In the real world of EIAs, not many of the environmental impacts that cause the greatest controversy are capable of being accurately quantified in a way that is meaningful. Simple deterministic calculations can be made on traffic movements (e.g. 100 vehicle miles per day) or dust emissions (cubic metres per day) and similar, but how do you quantify the visual impact of a pipeline installed through a forest, for example? It all depends on whether you happen to be there to see it. What is a visual eyesore to some becomes an engineering tourist attraction to others.

To try and bring some useful comparisons to this, a family of designations is used, with a ranking number to give them a measure of respectability. The ecological impact of clearing a strip of land occupied by a particular species of small newt or lizard would then have to be classified as something like one of the following categories.

0: No impact: acceptable
1: No desirable impact
2: Low impact
3: Medium impact
4: Medium-to-high impact
5: Irreversible impact

On top of this, because very little in the world is certain, EIA conclusions are inevitably *probabilistic* rather than deterministic. The happenings in the previous table must thus be multiplied by a percentage probability of each occurring, such as

Zero probability: 0%
Negligible probability: 0–1%
Low probability: 1–20%
Medium probability: 20–50%
Medium–high probability: 50–70%
High probability: 70–99%
Certainty: 100%

Hence the conclusion of an EIA statement might read something like

> Mitigation measures taken to reduce wind-erosion of cleared areas will result in a low probability of a dust emission factor of 1200 kg/acre-month.

The definitions of the terms are included in the EIA, but the fact remains that definitions of low, medium, high etc. and their attendant probability figures can be chosen more or less as required. You can now see the problem with interpreting and comparing EIA impact rankings, if that was something you had to do.

The role of the PI in EIAs

There isn't much of a role for the PI in the production of an EIA. The EIA is written long before a construction PI gets involved with a project and, unless called to provide information on construction activities (unlikely), the PI plays no part. However, there is a role in verifying any agreed environmental *mitigation measures* both during the construction and during the reinstatement of the pipeline RoW after it is completed. The agreed measures should have been incorporated into the contractor's project procedures but it is not exactly unknown for them to be forgotten, having been decided many years ago. Remember that, if the API 1169 exam breakdown is anything to go by, then 15% of the role of the PI is devoted to pipeline construction environmental protection.

18.10 How to learn from codes

Technical codes do not rank highly on the scale of exciting documents to read. They consist mainly of heavy, densely spaced text, often littered with multiple cross-references. Whereas design codes have a logical order to them, post-construction codes (PCCs) and recommended practice (RP) documents have a less firm structure. US codes, in particular, adopt a looser style, containing thousands of information snippets rather than building arguments in a logically developed step-by-step way. This makes them difficult to assimilate unless you have been brought up to rely on them for day-to-day guidance.

You have a problem

The problem you have is that your brain doesn't like information that is presented in the form of thousands of information snippets. It prefers things that are in order or that build into an assembly of facts that relate to each other, the more neatly the better.

The problem is, without the idealised structure that it likes, your mind's attention span for diverse, loosely connected information is a non-impressive

about 60 seconds

It will understand and absorb information within this 60-second window, until it tires of the game and starts to think about something else. You won't notice this happening, you will continue to drift through reading it, but will understand and retain almost none of it, until you eventually lose interest and switch off. It is possible to revitalise your interest by consciously refocusing, but this can only last for so long until your concentration lapses once again. The battle is lost.

Would you like to solve this problem?

Your 60-second attention limit is easy to solve – all you have to do is to help your brain work in the way it likes to. The brain is, by design, a problem-solving machine – given a question it will search for the answer on several subconscious levels that you know nothing about. Once it has found the answer, it will not only remember it, but will remember *how* it found it, and be able to apply the same process to similar or related problems.

So, if you ask yourself a question and then read a code section with the objective of *finding the answer*, your attention span will increase dramatically as your subconscious looks for the answer. Once it finds it, it reloads, ready to start again.

Once again, here is the way to do it.

1. Look at the title and contents of a code section.
2. Ask yourself a question *that you would like the section to answer for you.*
3. **Now** read through all of the section looking for the answer to your question.

With practice you can do this for up to four or five questions in parallel, reducing the number of times you need to read through technical documents to obtain a reasonable understanding of their relevant content.

What barriers stand in your way?
Only those that you put in place yourself. There are three.

- Barrier 1: you aren't really interested. If you don't have a genuine interest in what is in a code section, then your mind will not bother searching for the answers to any half-hearted questions you ask it. You will get little reward – so do yourself a favour and go do something else.
- Barrier 2: your questions are imprecise. Ask yourself a general, low-

stress question and you will soon find a general, equally low-stress answer that will keep you happy. Unfortunately, you have been fooled – effective learning happens in high-stress environments driven by necessity and urgency, not in your comfort zone. Stay in your comfort zone and you will learn at glacial speed.
- Barrier 3: you give in too easily. It is only too easy to give up searching for answers in code clauses. They are, after all, not that exciting. Why not wait for someone else to find the answer for you? Maybe they will if you wait long enough. Congratulations, you have just reinvented barrier 1.

Using this method in practice

Expect this to be a high-stress, tiring occupation. It is evidence that your accelerated learning is working. You are finding new information, processing and storing it away, then returning to find more. This carries an energy cost, and the more you learn the more tiring it is.

It is virtually impossible to learn at this accelerated rate for extended periods. The optimum time is

about 45 minutes

This defines the best session length to use during your code-learning activities. Sessions of less than 30 minutes are not really long enough to obtain continuity of learning, whereas any longer than 45 minutes will result in a quick fall-off in your learning efficiency. You also need to leave at least 4 hours between these sessions to give your concentration time to recover.

Try it out.

Chapter 19

Mock examination: 100 questions

Q1

Dry ditch crossing methods for crossing of waterbodies up to 30 feet wide mentioned in FERC wetland and waterbody procedures are dam and pump, flume method and

(a) Closed-cut method
(b) Directional blasting method
(c) HDD
(d) Geodesic bridging

Q2

According to OSHA, caution signs are used to warn against

(a) Danger
(b) All types of hazards
(c) Safety instruction signs
(d) Potential hazards only

Q3

If a substantial leak is discovered during hydraulic pressure testing of a pipeline then

(a) The test should be stopped before full pressure is achieved
(b) The leak should be located and the test continued to completion
(c) The test should be discontinued immediately then repairs carried out
(d) The pressure should be reduced while locating the leak, then the test discontinued and repairs carried out

Q4

According to OSHA, when undergoing a QNFT respirator fit test, an employee wearing the respirator must be able to do normal and deep breathing, turn their head from side-to-side, up and down and

(a) Frown, grimace (make funny faces) ☐
(b) See clearly in a 180 degree arc ☐
(c) Jog on the spot ☐
(d) Touch their toes ☐

Q5

According to OSHA lockout/tagout procedure requirements, employers shall ensure that training in the recognition of applicable hazardous energy sources, the type and magnitude of the energy available in the workplace and the methods and means necessary for energy isolation and control is provided to

(a) Each employee in the area ☐
(b) Each authorised employee ☐
(c) Each employee mentioned on the work permit ☐
(d) All employees ☐

Q6

When does a pipeline construction inspector have 'stop work' authority?

(a) Never; the role is limited to monitoring and reporting ☐
(b) Whenever a task or test has been done incorrectly ☐
(c) When there is imminent danger to people or environment ☐
(d) Only when the owner or site Health & Safety representative agrees ☐

Q7

After land reinstatement has been completed after a pipeline installation in an upland area, for how long is the project sponsor obliged, in co-operation with the landowners, to control the unauthorised use of off-road vehicles around the RoW site?

(a) Minimum 2 years ☐
(b) Minimum 4 years ☐
(c) Minimum 10 years ☐
(d) For the life of the project ☐

Q8

According to OSHA, horizontal members used along the walls of an excavated trench as part of an aluminium hydraulic shoring system to prevent cave-in are called

(a) Wales
(b) Walls
(c) Wiles
(d) Rails

Q9

During the lowering-in of a pipeline into its excavated trench, the pipeline construction inspector should ensure that side-bends are kept clear of

(a) The upper bench level
(b) The trench bottom
(c) The trench wall
(d) Stiffening braces

Q10

Which of the following statements relating to pipeline RoW preparation is true, according to API 1169?

(a) Pipeline inspectors can overrule the landowner's requirements
(b) The pipeline inspector acts as the owner/user's main contact with the landowner
(c) There is usually interaction between the pipeline inspector and the landowners
(d) It is not the pipeline inspector's role to interact with the landowner

Q11

OSHA 29 CFR 1910 covers requirements for minimising or preventing the consequences of catastrophic release of chemicals which are toxic, flammable, explosive or

(a) Catalytic
(b) Reactive
(c) Endothermic
(d) Exothermic

Q12

API 1169 is a

(a) Construction standard ☐
(b) Recommended Practice ☐
(c) Code ☐
(d) Mixture of all the above depending on the jurisdiction ☐

Q13

What is the role of the pipeline inspector during pipeline surveys performed prior to forest or other ground clearing?

(a) Liaising with national/jurisdiction authorities ☐
(b) Deciding RoW limits ☐
(c) Monitoring workmanship ☐
(d) None ☐

Q14

Inspectors designated to the API 1169 'Chief Inspector' role should

(a) Have at least 10 years' experience in pipeline projects ☐
(b) Be qualified to PE (Professional Engineer: US(A) level ☐
(c) Have at least 5 years' experience in pipeline projects ☐
(d) Have served in several different inspection classifications ☐

Q15

Which of these statements is true about a pipeline pressure test performed to RP 1110?

(a) Temperature is easier to determine accurately than pressure ☐
(b) Temperature is more difficult to determine accurately than pressure ☐
(c) Temperature and pressure are unrelated ☐
(d) Low test temperature can result in pressure reversal ☐

Q16

FERC upland maintenance plan requires that when using hay or wood-fibre mulch as an erosion control measure

(a) It may be applied before or after seeding ☐
(b) It must be applied immediately after seeding ☐
(c) It shall be used as a temporary, not permanent measure ☐
(d) It must not be applied within 100 feet of a wetland or waterbody ☐

Q17

If unwanted timber is burnt during clearing of a pipeline RoW, the burn locations should

(a) Not be on a solid area such as a temporary car park ☐
(b) Be carried out near a watercourse to dispose of ash ☐
(c) Have location recorded using GPS co-ordinates ☐
(d) Only burnt during daytime hours so the fire can be controlled ☐

Q18

On a pipeline RoW, the interface between land and a river or stream is called the

(a) Margin zone ☐
(b) Cadastral boundary ☐
(c) Medial zone ☐
(d) Riparian zone ☐

Q19

What is the maximum permitted horizontal spacing of the hydraulic cylinders when aluminium hydraulic shoring is used to shore a 12-foot-deep trench excavated in type A soil?

(a) 5 feet ☐
(b) 8 feet ☐
(c) 10 feet ☐
(d) 20 feet ☐

Q20

Rip-rap, corduroy and rollback are terms used for

(a) Types of backfill for trenches ☐
(b) Types of liners used in trenches ☐
(c) Scrap timber used for pathways ☐
(d) Types of fences used to protect Pipeline RoWs from trespassers ☐

Q21

What is the maximum allowable distance between CP test stations on a long pipeline?

(a) 3 km ☐
(b) 3 miles ☐
(c) 5 miles ☐
(d) 5 km ☐

Q22

An API 1169 pipeline construction inspector may

(a) Help formulate regulatory requirements ☐
(b) Be asked to assist welding inspectors ☐
(c) Act as the owner's representative for contract financial matters ☐
(d) Not get involved with welders ☐

Q23

According to OSHA, the slope of a trench excavation face is expressed as

(a) The vertical rise in feet per 100 feet of trench length e.g. 1 in 100 would be 1% ☐
(b) The ratio of transverse horizontal distance (H) to vertical rise (V) of the trench side wall e.g. H:V ☐
(c) The ratio of vertical rise (V) to transverse horizontal distance (H) of the trench side wall e.g. V:H ☐
(d) The angle that the cut face of an excavated trench makes with the vertical e.g. 15° (degrees) ☐

Q24

Immediately before lowering coated pipeline lengths into their trench, it should be subject to inspection by

(a) Visual examination and alignment check
(b) Holiday testing
(c) Coating adhesion tester
(d) Pressure test

Q25

According to 49 CFR 195, production weld indications are assessed to the acceptance criteria in

(a) ASME V and IX
(b) ASNT-TC-1A
(c) API 1104
(d) ASME B31.8

Q26

When a pipeline girth weld is to be subject to RT under the requirements of 49 CFR 195, then

(a) The full circumferential weld shall be tested
(b) At least 4 films shall be shot
(c) At least a 6 in in length shall be tested
(d) Only the 12 in centred on the weld stop/start location is required

Q27

For pipeline construction in soil in non-urban areas, excavation of the pipeline ditch is typically performed after stringing and field bending, then

(a) Before welding, NDE and coating
(b) Before coating but after welding and NDE
(c) After welding, NDE and coating
(d) After welding and NDE before coating

Q28

In B31.4, a defect

(a) Is the same as an imperfection ☐
(b) Will reduce the design life of a component ☐
(c) Requires comparison with code acceptance limits ☐
(d) Is always rejectable ☐

Q29

A pressure test used to verify the structural integrity of a pipeline containing time-dependent anomalies is a

(a) Strength test ☐
(b) Spike test ☐
(c) Leak test ☐
(d) All of the above ☐

Q30

Corrosion control inspectors working on pipeline projects should be knowledgeable of corrosion control subjects and have completed

(a) API 571 certificate exam ☐
(b) NACE CIP Level 1 ☐
(c) SSPC Level 1 ☐
(d) API 1169 certificate exam ☐

Q31

When reviewing pipeline project drawings, the pipeline inspector should expect to see physical changes in layout etc. marked up on the drawings as

(a) A concession reference number starting with XX ☐
(b) Circle or bubble-shaped mark-ups ☐
(c) Red line mark-ups ☐
(d) Yellow highlighter pen mark-ups ☐

Q32

An OSHA *quantitative fit test* relates to

(a) The leakage of respirator equipment
(b) The physical fitness of an employee
(c) The fit of protective headgear
(d) None of the above

Q33

During preparation for pipeline installation, laying the pipeline sections and supports/skids end-to-end along the RoW is called

(a) Alignment checking
(b) Bunching or banding
(c) Spooling
(d) Stringing

Q34

If a pipeline inspector discovers work that does not meet one of the *non-technical* contract terms of a pipeline project, API 1169 requires that they

(a) Reject the work
(b) Report it to their superior
(c) Make reasonable attempts to have it corrected
(d) Only report it if it is of a technical or QA/QC nature

Q35

In API 1169, the abbreviation ROW means

(a) Risk of waterlogging
(b) Right of way
(c) Recommended operating window
(d) Remote operated winch

Q36

A pipeline strength pressure test is considered successful if

(a) No ruptures or leaks occur
(b) No elastic deformation can be measured
(c) No pressure drops occur
(d) All of the above

Q37

During pipeline construction, the CEPA/INGAA states that

(a) A non-conformance may be a one-off deficiency ☐
(b) An identified deficiency needs to be corrected ☐
(c) A non-conformance is an isolated deviation from requirements ☐
(d) None of the above ☐

Q38

In API 1169, horizontal directional drilling (HDD) inspectors report to the

(a) HDD company management ☐
(b) Pipeline contractor ☐
(c) Owner/user ☐
(d) None of the above ☐

Q39

A thorough inspection of alloy steel chain slings shall include a thorough inspection for wear, defective welds, increase in length and

(a) Shredding ☐
(b) Creep ☐
(c) Deformation ☐
(d) Brittle fracture ☐

Q40

According to OSHA, piles of soil or other bulk material that may be at risk of falling or rolling into an excavated trench containing personnel shall

(a) Not be more than 24 in high ☐
(b) Be kept at least 2 feet from the edge of the excavation ☐
(c) Be kept at least 10 feet from the edge of the excavation ☐
(d) Be clearly marked with 'danger' signs 2 feet from the edge ☐

Q41

When installing a gas pipeline crossing to INGAA guidelines, an excavation tolerance zone is an area near an existing pipeline facility within which

(a) Soil must be removed by approved mechanical means rather than hand digging by external contractors ☐
(b) Soil must be removed by non-mechanical means ☐
(c) Soil must not be removed before authorization by the landowner ☐
(d) Soil must not be removed at all ☐

Q42

The objective of the FERC wetland and waterbody procedures document is to

(a) Prevent pipelines crossing them ☐
(b) Prevent pipelines being run parallel to them ☐
(c) Allocate environmental mitigation procedures ☐
(d) Minimise disturbances to them ☐

Q43

An instrument used for automatically maintaining a controlled pressure on the system when a pipeline pressure test is in progress is a

(a) Deadweight tester ☐
(b) Accumulation system ☐
(c) Spike tester ☐
(d) Surge arrestor ☐

Q44

According to OSHA, if a crane operator is unsure of the weight of a proposed load and no weight calculations are available then they

(a) Must 'quarantine' the load with signs and report this to their employer ☐
(b) Must not commence the lift ☐
(c) May do a quick test by trying to pull the load sideways without lifting it. If it moves, then the lift can go ahead ☐
(d) May start the lift using a load weighing device and as long as it doesn't read $>75\%$ SWL, proceed with the lift ☐

Q45

A pipeline inspector involved in site coating during pipeline construction

(a) May perform spark (Holiday) testing and thickness testing only ☐
(b) May not perform coating activities on their own ☐
(c) Should also be qualified in coating application procedures ☐
(d) Is expected to undertake significant coating inspection ☐

Q46

An arc welding process for pipelines that does not require manual manipulation of the arc but may require operator interruption in guiding or tracking the weld direction is what type of welding process?

(a) Manual ☐
(b) Semi-automatic ☐
(c) Autogenous ☐
(d) Automatic ☐

Q47

According to OSHA, a person capable of identifying existing and prohibited hazards in working conditions and who has authorisation to take prompt corrective action to eliminate them is

(a) An authorised person ☐
(b) An approved person ☐
(c) A designated person ☐
(d) A competent person ☐

Q48

Who is formally responsible for all safety on a pipeline work site?

(a) The contractor ☐
(b) The owner/user ☐
(c) The governing authority or jurisdiction ☐
(d) Everyone involved ☐

Q49

49 CFR 195 suggests that pre-1970 electrical resistance welded (ERW) or lap-welded pipe

(a) Requires additional pressure testing ☐
(b) Is unlikely to be at risk of longitudinal seam failure ☐
(c) Requires 100% NDE ☐
(d) Is at risk of longitudinal seam failure until proven otherwise ☐

Q50

According to OSHA soil classification, a moist cohesive soil is

(a) A soil that exhibits open cracks in an exposed surface ☐
(b) A soil which is free-seeping (or may be underwater) ☐
(c) A soil that can be rolled into small diameter threads before crumbling ☐
(d) A soil in which the voids are filled with water ☐

Q51

Dry ditch crossing methods for crossing of waterbodies up to 30 feet wide mentioned in FERC wetland and waterbody procedures are dam and pump, flume method and

(a) OCM (open cut method) ☐
(b) DB ☐
(c) GDP ☐
(d) HDD ☐

Q52

When regulatory (state or federal) bodies are involved in inspection/visits to pipeline construction sites the pipeline inspectors should have sufficient knowledge to enable them to

(a) Provide the agency with the information it requests ☐
(b) Determine the agency involved and their objectives ☐
(c) Write a site procedure to deal with such events ☐
(d) Interpret the regulations on behalf of the owner/user ☐

Q53

In a B31.8 gas pipeline system, gas may be treated to reduce its corrosivity by

(a) Filtering ☐
(b) Ozoning ☐
(c) Sweetening ☐
(d) Souring ☐

Q54

Which organisation shall apply for FERC authorisation for a pipeline RoW planned to run across or near bodies of water or wetlands?

(a) The project sponsor organisation ☐
(b) The excavation organisation ☐
(c) The relevant landowner in the RoW area ☐
(d) Any of the above, as appropriate ☐

Q55

In API Q1 the term MAC means

(a) Manufacturing Assurance Criteria ☐
(b) Manufacturing Acceptance Criteria ☐
(c) Manufacturing Activity Change ☐
(d) Management After Configuration ☐

Q56

Pipeline inspectors should check the types and sizes of flanged fittings against the code requirements of

(a) API 1104 ☐
(b) API 5L ☐
(c) ASME/ANSI B16.34 ☐
(d) ASME/ANSI B16.5 ☐

Q57

API 1169 requires that mainline pipeline welding shall be monitored and assessed by

(a) Any qualified individual to the requirements of CWI ☐
(b) The welding inspector to the requirements of ASME IX ☐
(c) The welding inspector to the requirements of API 1104 ☐
(d) The pipeline inspector to the requirements of API 1104 ☐

Q58

In ISO 9000, a part of quality management specifying operational processes and related resources to achieve quality objectives is

(a) Change control ☐
(b) Quality planning ☐
(c) Quality improvement ☐
(d) Quality assurance ☐

Q59

When two pipeline routes cross, the crossing angle should be

(a) As near to 90° as practical ☐
(b) As near to 45° as practical ☐
(c) Located in an area of flat subsoil ☐
(d) Sufficient to align the cathodic protection (CP) potentials ☐

Q60

When welding pipelines to API 1169, a finding obtained by NDE is

(a) An indication ☐
(b) A defect ☐
(c) A flaw ☐
(d) An imperfection ☐

Q61

If a fire is discovered in an area where blasting explosives are stored then

(a) All employees shall be immediately supplied with respirators ☐
(b) All adjacent areas shall, as a precaution, be flooded with foam to a depth of at least 3 feet ☐
(c) The site employee fire-fighting plan should be put into effect ☐
(d) Employees should not fight the fire ☐

Q62

Under API Q1 product realization procedures, a non-conforming manufactured product that does not satisfy manufacturing acceptance criteria (MAC) shall

(a) Be permitted to be released if the customer agrees to a concession ☐
(b) Not be permitted to be released to the customer ☐
(c) Be remanufactured or repaired in all cases ☐
(d) Be permitted to be released to the customer if the manufacturing organisation's top management approve it ☐

Q63

Which of these is not covered by the scope of RP 1110 *Pressure Testing of Pipelines*?

(a) Compressor station piping smaller than NPS 10 ☐
(b) Pumping units ☐
(c) Pump station piping ☐
(d) Pipeline appurtenances (components attached to the pipeline) ☐

Q64

According to OSHA, procedures related to the lifting capacity of cranes must be

a) Developed by a qualified person but who does not necessarily need to be a Professional Engineer (PE) ☐
(b) Approved by all involved construction site personnel before being implemented ☐
(c) Developed and signed off by a PE familiar with the equipment ☐
(d) Developed by a PE employed by the manufacturer and signed off by a qualified person who will be in charge of lifting operations on the construction site itself ☐

Q65

When repairing welds on pipe spools being welded to API 1104 repair, then weld repair procedures

(a) Shall be qualified by destructive testing only ☐
(b) Are required at the discretion of the company ☐
(c) Shall be qualified by visual and destructive testing ☐
(d) Shall be qualified by visual and destructive testing and NDE ☐

Q66

NDE personnel involved in pipeline construction

(a) May be qualified to any national certification programme acceptable to the company ☐
(b) Shall be qualified to ASNT-TC-IA ☐
(c) Must be ASNT-TC-IA level 3-qualified to interpret test results ☐
(d) Only level 1-qualified personnel shall interpret NDE results ☐

Q67

When RT is performed on a pipeline circumferentially welded to API 1104 using a single exposure with the radioactive source inside the piping (i.e. a panoramic shot), what is the minimum number of source-side IQIs required to be placed around the circumference?

(a) A single IQI is adequate ☐
(b) 2 on the inside and 2 on the outside ☐
(c) 4 on the inside ☐
(d) 4 on the outside or inside ☐

Q68

When does a pipeline construction inspector have 'stop work' authority?

(a) Whenever a task or test has been done incorrectly ☐
(b) Only when the owner or site Health & Safety representative agrees ☐
(c) Only when there is imminent danger to people or environment ☐
(d) Never; the role is limited to monitoring and reporting ☐

Q69

According to the CEPA/INGAA guide, which of these is the responsibility of the inspector during pipeline construction?

(a) Acting as mentor to less experienced workers ☐
(b) Helping people select the correct tools for jobs ☐
(c) Ensuring a drugs/alcohol policy is in place ☐
(d) Reporting people for not wearing seatbelts in their cars ☐

Q70

When witnessing a hydrostatic pressure test on a pipeline, the pipeline inspector should verify the test results, report on any pressure variations and

(a) Choose the test pressure ☐
(b) Choose the test medium ☐
(c) Report temperature variations ☐
(d) Define the role of the owner/user ☐

Q71

What is the purpose of roller dollies when roll welding circumferentially weld sections of pipeline?

(a) Chamfering the joint preparation ☐
(b) Prevent sagging ☐
(c) Guiding the welding head ☐
(d) Maintaining alignment of the joint fit-up ☐

Q72

When are pipeline inspectors empowered and expected to shut down a pipeline work activity before it has started?

(a) Never, it is not their job ☐
(b) Only when the owner/user agrees ☐
(c) When their employer instructs them ☐
(d) When imminent danger to property is anticipated if work goes ahead ☐

Q73

The code covering *Specification for Line Pipe* is

(a) API 1104
(b) API 5L
(c) API 6D
(d) ASME B13.8

Q74

Pipeline survey involves

(a) Defining RoW limits
(b) Monitoring of materials arriving at the site
(c) Regular inspections of the pipeline route in use
(d) All of the above

Q75

Under 49 CFR 192, NPS 24 steel gas pipelines operating at a stress <20% SMYS must have what size of dents removed?

(a) None, they can all remain in place
(b) Those with any sharp edges or near longitudinal or circumferential welds
(c) With depth $> \frac{1}{4}$ in (6.4 mm)
(d) With depth $> 2\%$ of nominal pipe diameter

Q76

The CEPA/INGAA good practice guide covers

(a) Hydrostatic testing only
(b) Hydrostatic and pneumatic testing
(c) Hydrostatic and spike testing
(d) All types of pressure testing

Q77

In addition to the 1169 certified pipeline inspector, API 1169 (annexes) describes the roles of the welding inspector, blasting inspector, horizontal drilling inspector and

(a) API 1169 certified pipeline inspector ☐
(b) Corrosion control inspector and chief inspector ☐
(c) Inspector co-ordinator and corrosion control inspector ☐
(d) NDE inspector (examiner) and coating inspector ☐

Q78

Compared to hydrostatic testing, pneumatic testing

(a) Is safer ☐
(b) Is preferred by CEPA/INGAA good practice guide ☐
(c) Is better at finding small leaks ☐
(d) Is quicker ☐

Q79

Which areas of a welded pipeline normally have their external coating applied on the construction site?

(a) All of them ☐
(b) Girth welds only ☐
(c) Girth welds and longitudinal welds only ☐
(d) Areas within 3 feet of a site weld ☐

Q80

In API Q1 'Product Realization' section, which activity immediately follows 'design and development outputs'?

(a) Design and development verification ☐
(b) Design and development validation and approval ☐
(c) Design and development inputs ☐
(d) Design and development review ☐

Q81

During the lowering-in of a pipeline into its excavated trench the main focus of a pipeline inspector should be to

(a) Check the backfill depth and compaction is correct ☐
(b) Ensure sufficient clearance to allow in-trench coating ☐
(c) Check the trench depth is adequate for minimum coverage requirements ☐
(d) Monitor coating integrity ☐

Q82

Cathodic protection (CP) of a pipeline acts, in a passive CP system, by making a sacrificial anode

(a) Anodic to the pipe ☐
(b) Cathodic to the pipe ☐
(c) Receive an impressed negative potential from a rectifier unit ☐
(d) Receive an impressed positive potential from a rectifier unit ☐

Q83

49 CFR 195 does not cover

(a) Hazardous liquid pipelines ☐
(b) Hazardous gas pipelines ☐
(c) Any liquid pipelines ☐
(d) Steel cross-country pipelines ☐

Q84

During preparation for pipeline installation, stockpiles of owner-supplied materials (e.g. pipe sections) should be stored

(a) Within the agreed RoW boundaries as long as the landowner agrees ☐
(b) Away from the agreed RoW boundaries ☐
(c) In security-controlled compounds ☐
(d) Under tarpaulins or other dry conditions ☐

Q85

Which of these is not a common safety policy/practice/procedure that would be expected to be put into place by the pipeline owner company?

(a) Vehicle safety practice
(b) Hearing conservation practice
(c) Working alone policy
(d) Environmental reporting practice

Q86

How should a pipeline inspector react if a concern is expressed to the pipeline inspector by a passing member of the public or an unconnected landowner?

(a) Suggest the enquirer contact the local authority or jurisdiction
(b) Attempt to provide an explanation
(c) Report the occurrence to the owner/user
(d) Ignore it – it is outside the pipeline inspector's responsibility

Q87

Cold bending on site of large diameter pipeline sections to make them fit properly

(a) Is not allowed
(b) Is a valid replacement for hot bending for small radius bends
(c) May only be done on pipeline spools that are seamless
(d) Is normal

Q88

The minimum soil cover over normally excavated buried pipeline in cultivated, residential or industrial areas is

(a) 24 in
(b) 36 in
(c) 48 in
(d) 72 in

Q89

According to the CEPA/INGAA best practice guide, an EPP on a site where no animals have been seen relates to

(a) Wildlife habitats
(b) Fire prevention
(c) Health and safety
(d) All emergency procedures

Q90

According to the CEPA/INGAA guide, when documentary records are found to be incomplete during or after a pipeline construction then

(a) It is a matter for the pipeline owner, not the construction inspector
(b) Construction inspector duties are considered incomplete
(c) The pipeline cannot be commissioned until they are completed
(d) It has to be reported to all parties in the RoW agreements

Q91

A pipeline construction inspector does not

(a) Monitor activities against regulatory requirements
(b) Monitor contractor progress
(c) Monitor activities against safety requirements
(d) Plan the contractor's progress

Q92

A segment of B31.8 pipeline installed in a gas transmission system or between storage fields is a

(a) Transmission line
(b) Transmission system
(c) Pipeline facility
(d) Pipeline section

Q93

When can flammable natural gas be used as a pressure test medium for a gas pipeline in class 2 location at a pressure of 1.25×MOP?

(a) If an exclusion zone of 1 mile is maintained around the test location ☐
(b) If the test pressure is <200 psi ☐
(c) If stress levels are kept below 3% SMYS ☐
(d) Never ☐

Q94

During pipeline project, the step immediately following RoW surveying is

(a) Soil analysis testing ☐
(b) Wildlife survey ☐
(c) Clearing and grading ☐
(d) Ditching and excavation ☐

Q95

According to OSHA, a person is considered to have entered a permit-controlled confined space when

(a) The top half of their body above the waist breaks the plane of an opening into the space ☐
(b) Their head breaks the plane of an opening into the space ☐
(c) Any part of their body breaks the plane of an opening into he space ☐
(d) All their body is inside the plane of an opening into the space ☐

Q96

'ONE CALL' is

(a) A documented risk assessment/JSA process ☐
(b) An arrangement for verbal site commands between personnel ☐
(c) A telephonic accident reporting system ☐
(d) A telephonic excavation notification system ☐

Q97

During pipeline construction, an isolated deviation from requirements that does not impact structural integrity, cost or construction schedule is defined in the CEPA/INGAA guide as

(a) A supplier observation (SO) ☐
(b) An identified deficiency ☐
(c) A non-conformance ☐
(d) Any of the above, depending on the construction inspector's viewpoint ☐

Q98

Pipeline construction site clean-up after installation

(a) Should be completed before the end of winter in forested areas ☐
(b) Should be completed in a maximum of 6 weeks in urban areas ☐
(c) May be performed in phases depending on location and season ☐
(d) Should be performed in a single series of actions ☐

Q99

According to API Q1, QMS documentation controls the quality management system processes and

(a) Identification of key point indicator (KPIs) ☐
(b) Identification of legal requirements that are needed to achieve product conformity ☐
(c) List of design acceptance criteria ☐
(d) List of registered designs, patients, copyrights etc. that, as part of the QMS, form an essential quality link with customers and hence the continued success of the business ☐

Q100

If a small leak is discovered during hydraulic pressure testing of a pipeline then

(a) The test should be stopped before full pressure is achieved ☐
(b) The leak should be located and the test continued to completion ☐
(c) The test should be discontinued immediately then repairs carried out ☐
(d) The pressure should be reduced while locating the leak, then the test discontinued and repairs carried out ☐

Appendix 1

Answers to sample questions

Question set 6.1: API RP 1169 – responsibilities

Question	Answer	API RP 1169
Q1	(b)	4.2
Q2	(b)	4.4
Q3	(a)	7.18.4
Q4	(c)	5.11.2
Q5	(b)	7.3.7
Q6	(b)	4.5
Q7	(a)	6.7
Q8	(c)	4.7
Q9	(d)	Annex B1
Q10	(b)	7.2.3
Q11	(d)	4.3
Q12	(c)	7.2.1
Q13	(c)	5.4.3
Q14	(d)	Annex A 2.3.1
Q15	(b)	Annex D.2

Question set 6.2: CEPA/INGAA guide – responsibilities

Question	Answer	CEPA/INGAA guide
Q1	(d)	6.1
Q2	(d)	6
Q3	(b)	Table 4
Q4	(d)	6.6
Q5	(d)	6.4.1
Q6	(b)	13.4
Q7	(c)	6.1
Q8	(a)	Table 29
Q9	(d)	6
Q10	(a)	Table 6

Question set 6.3: INGAA/CEPA guide – general

Question	Answer	INGAA/CEPA guide
Q1	(c)	Table 59
Q2	(a)	6.2 Table 2
Q3	(b)	6.3 Table 4
Q4	(c)	10.6 Tables 62 & 65
Q5	(a)	6.8 Tables 10
Q6	(a)	6.4.2 Table 6
Q7	(a)	11.3 Tables 88 & 89
Q8	(c)	9.3 Tables 48 & 55
Q9	(d)	11.3 Table 80
Q10	(d)	14.5 Table 100

Question set 6.4: CEPA/INGAA guide – inputs and outputs

Question	Answer	INGAA/CEPA guide
Q1	(d)	Restoration (18.1)
Q2	(c)	Pressure Testing Table 151
Q3	(b)	Stringing Table 58
Q4	(b)	Backfilling Table 115
Q5	(b)	Survey Table 27
Q6	(d)	Ditching Table 81
Q7	(a)	Welding Table 84
Q8	(a)	Lowering-in Table 100
Q9	(d)	Coating Table 97
Q10	(c)	Survey Table 14
Q11	(c)	Backfilling Table 115
Q12	(b)	Cathodic Protection Table 131
Q13	(a)	Clearing Table 46
Q14	(a)	Restoration Table 166
Q15	(c)	Ditching Table 69
Q16	(c)	Welding Table 89
Q17	(b)	Stockpiling Table 49
Q18	(d)	Clearing Table 30
Q19	(b)	Field Bending Table 66
Q20	(c)	Survey Table 14

Answers to sample questions

Question set 6.5: CEPA/INGAA guide – monitoring activities

Question	Answer	INGAA/CEPA guide
Q1	(c)	Cathodic Protection Table 129
Q2	(c)	Field bending Table 65
Q3	(b)	Backfilling Table 116
Q4	(d)	Lowering in Table 105
Q5	(d)	Lowering in Table 102
Q6	(a)	Backfilling Table 116
Q7	(a)	Lowering in (14.2)
Q8	(d)	Ditching Table 71
Q9	(a)	Cathodic Protection Table 129
Q10	(d)	Stringing Table 54
Q11	(c)	Stringing Table 54
Q12	(a)	Stringing Table 51
Q13	(b)	Lowering in Table 105
Q14	(b)	Stringing Table 53
Q15	(b)	Coating Table 96

Question set 9.1: API 1104 pipeline welding

Question	Answer	API 1104
Q1	(a)	6.1
Q2	(a)	8.1
Q3	(b)	8.4.2
Q4	(b)	3.1.28
Q5	(a)	3.1.26
Q6	(a)	5.4
Q7	(b)	5.1
Q8	(b)	3.1.11
Q9	(d)	9.6.1.3
Q10	(c)	9.3.3

Question set 11.1: RP 1110 pipeline pressure testing

Question	Answer	RP 1110
Q1	(b)	RP 1169 (7.18.4)
Q2	(d)	RP 1110 (3.1.35)
Q3	(c)	General
Q4	(c)	RP 1110 (4.1.7.4)
Q5	(a)	RP 1110 (10.4)

Q6	(b)	RP 1110 (5.3)
Q7	(a)	RP 1110 (5.6b)
Q8	(a)	RP 1110 (4.1.11)
Q9	(a)	RP 1104 (5.5)
Q10	(c)	RP 1110 (4.1.2a) and 4.1.7.2

Question set 13.1: CGA best practice document (13.0)

Question	Answer	CGA 13.0
Q1	(a)	9
Q2	(a)	8.2
Q3	(c)	5.1
Q4	(d)	7.5
Q5	(c)	App A
Q6	(b)	Introduction
Q7	(c)	3.18
Q8	(b)	3.1
Q9	(a)	4.2
Q10	(c)	7.1
Q11	(a)	Page 1
Q12	(c)	App A

Question set 13.2: INGAA pipeline crossing guidelines

Question	Answer	INGAA pipeline crossing guidelines
Q1	(c)	II(m)
Q2	(c)	II(c)
Q3	(c)	Sec 1
Q4	(a)	Sec 1
Q5	(b)	II(f)
Q6	(a)	III guidelines
Q7	(d)	II(d)
Q8	(a)	II(m)
Q9	(b)	II(k)
Q10	(c)	II(i)

Answers to sample questions

Question set 15.1: OSHA 29 CFR 1910

Question	Answer	OSHA 29 CFR 1910
Q1	(d)	App C
Q2	(d)	1910.147
Q3	(d)	1910.147
Q4	(c)	1910.134
Q5	(c)	Page 7
Q6	(b)	1910.148
Q7	(a)	1910.184
Q8	(b)	1910.145
Q9	(c)	Page 12
Q10	(a)	1910.136

Question set 15.2: OSHA 29 CFR 1926

Question	Answer	OSHA 29 CFR 1926
Q1	(d)	Annex A
Q2	(c)	1926.451
Q3	(c)	1926.152 (5c)
Q4	(b)	1926.651
Q5	(b)	Annex B
Q6	(a)	1926.1417 (j1)
Q7	(b)	1926.1417y
Q8	(a)	1926.251 (a6)
Q9	(c)	1926.451
Q10	(b)	1926.62f
Q11	(b)	1926.65
Q12	(d)	1926.251
Q13	(c)	1926.353(c2)
Q14	(c)	1926.1417(b2)
Q15	(a)	1926.62

Question set 15.3: 49 CFR 172 hazardous materials table

Question	Answer	49 CFR 172 hazardous materials table
Q1	(c)	Page 54
Q2	(b)	Page 145
Q3	(a)	Page 45
Q4	(b)	Page 7
Q5	(d)	Page 6

Question set 15.4: ANSI Z49.1: safety in welding & cutting

Question	Answer	ANSI Z49.1
Q1	(a)	E4.3
Q2	(c)	4.1.3/4.1.4
Q3	(c)	E4.2.1.1
Q4	(c)	E6.3
Q5	(a)	8.6.1
Q6	(b)	5.5.3
Q7	(c)	E4.2.1.1
Q8	(d)	4.5
Q9	(b)	5.2
Q10	(b)	6.2.4

Question set 16.1: 40 CFR 300

Question	Answer	40 CFR 300
Q1	(b)	Page 17
Q2	(b)	Page 11
Q3	(a)	Page 20
Q4	(c)	300.405b
Q5	(d)	300.430a
Q6	(c)	300.425b
Q7	(a)	300.400b
Q8	(d)	300.5
Q9	(c)	300.1
Q10	(b)	300.425b

Question set 16.2: Navigable waterways: 33 CFR 321/33 USC Chapter 9

Question	Answer	33 CFR 321 and 33 USC Chapter 9
Q1	(c)	33 CFR 321 (321.2b)
Q2	(b)	33 CFR 321
Q3	(b)	33 CFR 321 (321.3b)
Q4	(d)	33 USC Ch9 (403a)
Q5	(a)	33 CFR 321 (321.3a)

Question set 16.3: Migratory bird permits: 50 CFR 21

Question	Answer	50 CFR 21
Q1	(b)	2.12 (8)
Q2	(a)	50 CFR 21
Q3	(a)	50 CFR 21
Q4	(c)	50 CFR 21
Q5	(d)	50 CFR 21

Question set 17.1

Question	Answer	Reference
Q1	(b)	API Q1 (5.11)
Q2	(c)	ISO 9000 (3.8.12)
Q3	(b)	RP1169 (4.6)
Q4	(a)	RP1169 (4.3)
Q5	(b)	ISO 9000 General principle
Q6	(c)	ISO 9000 (3.6.13)
Q7	(b)	ISO 9000 (4.5)
Q8	(c)	API Q1 (5.11)
Q9	(a)	API Q1 (4.4.3)
Q10	(d)	API Q1 (5.10)
Q11	(c)	RP1169 (4.7)
Q12	(b)	API Q1 (4.5)
Q13	(a)	API Q1 (8.4.2)
Q14	(c)	General knowledge
Q15	(b)	API Q1 (6.4.2)
Q16	(c)	ISO 9000 (2.3.4.3)
Q17	(d)	API Q1 (5.8)
Q18	(c)	API Q1 (5.8)
Q19	(a)	API Q1 (5.7.6.1)
Q20	(b)	API 578 (not in BoK)

Chapter 19 Mock examination: 100 questions

Question	Answer	Reference
Q1	(c)	FERC wetland and waterbody procedures
Q2	(d)	OSHA 29 CFR (1910.145)
Q3	(c)	RP 1110 (5.6a)
Q4	(a)	OSHA 29 CFR 1910 PPE
Q5	(b)	OSHA 29 CFR (1910.147)

Q6	(c)	CEPA Guide (6.1 Authorities)
Q7	(d)	FERC upland maintenance plan
Q8	(a)	OSHA CFR 1926 Subpart P (1926.650)
Q9	(c)	CEPA Guide (Table 105)
Q10	(c)	API 1169 (7.5.4)
Q11	(b)	OSHA 29 CFR (1910.119)
Q12	(b)	API 1169
Q13	(c)	CEPA Guide (7.3-4)
Q14	(d)	API 1169 (Annex A1)
Q15	(b)	RP 1110 (4.4)
Q16	(a)	FERC upland maintenance plan
Q17	(c)	CEPA Guide (Table 39)
Q18	(d)	CEPA Guide (7.5 Table 20)
Q19	(b)	OSHA CFR 1926 (App C Table D1-1)
Q20	(c)	CEPA Guide (8.1)
Q21	(d)	CEPA Guide (16.1)
Q22	(b)	CEPA Guide (6.0)
Q23	(b)	OSHA CFR 1926 (App B)
Q24	(b)	API 1169 (7.13.4)
Q25	(c)	49 CFR 195 (D195.228b)
Q26	(d)	49 CFR 195 (D195.234d)
Q27	(c)	CEPA Guide (11.1)
Q28	(d)	B31.4 (Def 400.2)
Q29	(b)	RP 1110 (4.1.7.2)
Q30	(b)	API 1169 (Annex E2.1)
Q31	(c)	API 1169 (4.8)
Q32	(b)	OSHA 29 CFR (1910.134)
Q33	(d)	CEPA Guide (9.1)
Q34	(a)	API 1169 (4.3)
Q35	(b)	API 1169 (3.2)
Q36	(a)	RP 1110 (4.1.7.3)
Q37	(d)	CEPA Guide (6.4.1 definitions)
Q38	(d)	API 1169 (Annex C.1)
Q39	(c)	OSHA 29 CFR (1910.148)
Q40	(b)	OSHA CFR (1926.651)
Q41	(b)	INGAA pipeline crossing guidelines (IIk)
Q42	(d)	FERC wetland/waterbody procedures (1A)
Q43	(a)	RP 1110 (def 3.1.10)
Q44	(d)	OSHA 29 CFR (1926.1417)
Q45	(b)	CEPA Guide (13.4)
Q46	(d)	API 1104 (Def 3.1.1)

Q47	(d)	OSHA 29 CFR (1926.29 Def)
Q48	(a)	API 1169 (5.2.3).
Q49	(d)	49 CFR 195 (E195.302d)
Q50	(c)	OSHA CFR 1926 (Annex A)
Q51	(d)	FERC wetland and waterbody procedures
Q52	(b)	API 1169 (5.18)
Q53	(c)	B31.8 (864.2.5)
Q54	(a)	FERC wetland and waterbody procedures 1A
Q55	(b)	API Q1 (Def 3.2)
Q56	(d)	API 1169 (7.8)
Q57	(c)	API 1169 (Annex D2.3)
Q58	(b)	ISO 9000 (Def 3.3.5)
Q59	(a)	INGAA crossing definition (IIc)
Q60	(a)	API 1104 (3.1.2)
Q61	(d)	OSHA CFR 1926 (Subpart U 1926.900f)
Q62	(a)	API Q1 (5.10.3)
Q63	(b)	RP 1110 (Sec 1 Scope)
Q64	(c)	OSHA 29 CFR (1926.1417b3)
Q65	(c)	API 1104 (10.3.3)
Q66	(a)	API 1104 (8.4.1)
Q67	(c)	API 1104 (11.1.6.1)
Q68	(c)	CEPA Guide (6.1 Authorities)
Q69	(d)	CEPA Guide 6.0 (Table 6)
Q70	(c)	API 1169 (7.18.4)
Q71	(b)	API 1104 (7.9.1)
Q72	(d)	API 1169 (4.6)
Q73	(b)	API 1169 (Sec 2: References)
Q74	(a)	CEPA Guide (7.1)
Q75	(a)	49 CFR 192 (G192.309)
Q76	(a)	CEPA Guide (17.1)
Q77	(b)	API 1169 (Annex A to E)
Q78	(c)	General knowledge
Q79	(b)	CEPA Guide (13.1)
Q80	(d)	API Q1 (5.4)
Q81	(d)	CEPA Guide (14.1)
Q82	(a)	CEPA Guide (16.1)
Q83	(b)	49 CFR 195 (D195.200)
Q84	(b)	CEPA Guide (9.1)
Q85	(d)	CEPA Guide 6.0 (Table 4)
Q86	(c)	API 1169 (4.9)
Q87	(d)	CEPA Guide (10.1)

Q88	(c)	B31.4 (Table 434.6.1)
Q89	(a)	CEPA Guide (7.5 Table 14)
Q90	(b)	CEPA Guide 6.8
Q91	(d)	CEPA Guide 6.0
Q92	(a)	B31.8 (803.2 definitions
Q93	(c)	49 CFR 192 (J192.503)
Q94	(c)	CEPA Guide (8.1)
Q95	(c)	OSHA 29 CFR (1910.146)
Q96	(d)	API 1169 (5.10.2)
Q97	(b)	CEPA Guide 6.0 (6.4.1 definition)
Q98	(c)	CEPA Guide (17.1)
Q99	(b)	API Q1 (4.4.1e)
Q100	(d)	RP 1110 (5.6a)

Appendix 2

API SIRE exam guide: sample chapter

This appendix presents a sample chapter from *Guide to Source Inspection (Rotating Equipment)* by the same authors (Figure A2.1). The book is divided into two parts.

Part A: Source inspection of rotating equipment
Chapter 1. How to use this book
Chapter 2. Source inspection: what's it all about?
Chapter 3. The role of the source inspector (SI)
Chapter 4. Tactics of source inspection (how to do it)
Chapter 5. Materials
Chapter 6. Oil systems and seals
Chapter 7. Centrifugal pumps
Chapter 8. Steam turbines
Chapter 9. Gas turbines
Chapter 10. Diesel engines
Chapter 11. Compressors (reciprocating, screw, rotary, axial)
Chapter 12. Power transmission

Part B: API SIRE exam preparation
Chapter 13. The API Individual Certification Programs (ICP)
Chapter 14. The API SIRE exam: what to expect
Chapter 15. The SIRE Body of Knowledge (BoK) and study guide
Chapter 16. Centrifugal pumps (API 610) and seals (API 682)
Chapter 17. GP steam turbines (API 611)
Chapter 18. Gear units (API 677) and LO systems (API 614)
Chapter 19. Compressors (API 617,618,619)
Chapter 20. Materials (API 578)
Chapter 21. NDE and welding (ASME V/IX)
Chapter 22. Surface preparation and painting (SSPC codes, MSS-SP-55)

The following is an abridged sample of the chapter concerning centrifugal pumps.

FIG A2.1

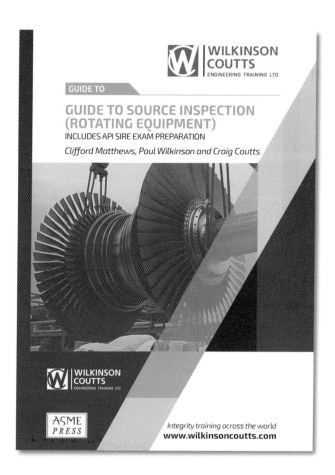

Centrifugal pumps

A2.1 Pump design

There are many different pump designs, but the centrifugal type is the most common in the process industry worldwide. Single-stage pumps are used for low-pressure, high-volume flow requirements, whereas multi-stage designs are used when higher pressure or total heads are required by the process. The static parts of a centrifugal pump comprise the pressure casing (stator) and its baseplate/mounting arrangement. From a source inspection viewpoint these are treated similarly to other cast or fabricated components with their requirement for material control, dimensional accuracy, NDE and pressure testing. The rotation (rotor) components comprise the forged shaft, cast impeller(s) and precision assemblies for bearings, seals, lubricating oil system and other small components.

Owing to the high rotational speed of the finished assembly, pumps have much tighter tolerances for dimensional, manufacturing and assembly accuracy than static pressure equipment. The performance of a pump (how much fluid it will pump over a range of suction and discharge pressures) is closely linked to its design, and the rotation brings with it the need for close control of dynamic balance and vibration limits.

Centrifugal pumps are not new technology, so their technical issues are well understood and incorporated into the technical specifications and standards that are established in the industry. All this is good news for the source inspector (SI); once you understand the basic principles of pump design and testing it is a fairly straightforward step to transfer this knowledge to other more complex types of rotating machinery.

A2.2 Pump performance criteria

The performance requirement of a pump is predominantly to do with its ability to move quantities of fluid. There are many pump performance parameters, some of which are complex and may be presented in a non-dimensional format. For source inspection purposes, however, it is only necessary to consider those that normally form part of the pump 'acceptance guarantees'.

Volume flow rate (q)

Flow rate is the first parameter specified by the process designer who bases the pump requirement on the flow rate that the process needs to function. This 'rated' flow rate is normally expressed in volume terms and it is represented by the symbol q, with units of cubic metres per second (m^3/s) or litres per second (l/s).

Head (H)

Once the rated flow rate has been determined, the designer then specifies a total head (H) required at this flow rate. This is expressed in metres or feet and represents the useable mechanical work transmitted to the fluid by the pump. Together, q and H define the *duty point*, which is the core performance criterion.

Net positive suction head (NPSH)

The NPSH is more difficult to understand. Essentially, it is a measure of the pump's ability to avoid cavitation in its inlet (suction) region. This is done by maintaining a pressure excess above the relevant vapour pressure in this inlet region. This pressure excess keeps the pressure above that at which cavitation will occur. Acceptance guarantees normally specify a maximum NPSH required. The unit is metres or feet.

Other performance criteria

Other performance criteria are

- pump efficiency (measured as a percentage value): the efficiency with which the pump transfers mechanical work to the fluid
- power (P) in watts (W) consumed by the pump
- noise and vibration characteristics.

It is normal practice for the above criteria to be expressed in the form of a set of 'acceptance guarantees' for the pump. The objective of a performance testing programme is therefore to demonstrate compliance with these guarantees.

FIG A2.2
Centrifugal pump characteristics

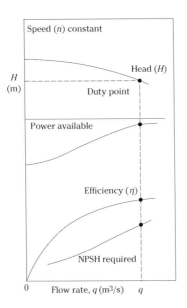

A2.3 Centrifugal pump testing

The set of curves that are used to describe pump performance are commonly known as the pump *characteristics*. Figure A2.2 shows a typical set.

The q/H curve

For most centrifugal pump designs the q/H characteristic looks like that shown in Figure A2.2. The test is carried out at a nominally constant speed and the head (H) decreases as flow rate (q) increases, giving a negative slope to the curve. Note how the required *duty point* is represented and how the required pump power and efficiency change as the flow rate varies.

The NPSH (required) curve

One reason why the NPSH can be confusing is because it needs two different sets of axes to describe it fully. The lower curve in Figure A2.2

shows how the NPSH *required* to maintain full head performance rises with increasing flow rate. Note that this curve is not obtained directly from the q/H test: it is made up of three or four points, each point being obtained from a separate NPSH test at a different constant q. This is normally carried out after the q/H test. The objective of the NPSH test is to demonstrate whether the pump can maintain full head performance at an NPSH *equal to or less than* a maximum guarantee value.

Typical acceptance guarantee schedule

Pump acceptance guarantees are expressed in quite precise terms. The example below shows indicative values for a large circulating water pump.

- Rated speed (n): 740 r/min.
- Rated flow rate (q): 0.9 m^3/s (the duty point).
- Rated total head (H): 60 m (the duty point).
- Rated efficiency: 80% at duty point.
- Absorbed power: 660 kW at duty point.
- NPSH: maximum 6 m at impeller eye for 3% total head drop.
- Vibration: maximum at the pump bearing = 2.8 mm/s root mean square (RMS) at duty point.
- Noise: maximum = 90 dB(A) at duty point (at agreed locations).
- Tolerances: ±1.5% on head (H) and ±2% on flow (q). These q/H tolerances are typical, but can be higher or lower, depending on what the designer wants. Tolerance is generally +0 on NPSH.
- The acceptance test standard: for example, API 610/ISO 13709. This is important – it tells you a lot about test conditions and which measurement tolerances to take into account when interpreting the performance curves.

A2.4 Pump specifications and standards

Pump performance testing is well covered by a tried and tested set of standards that relate specifically to the radial, mixed and axial flow category. These standards relate only to the pump itself. Pumps are only rarely subject to performance tests in the process system for which they are intended; they are normally tested in a specific performance test rig. The main standards are as follows.

- BS EN ISO 9906 (based on withdrawn standard ISO 3555) specifies three levels of testing tolerances:

- grades 1B, 1E and 1U with tighter tolerance
- grades 2B and 2U with broader tolerance
- grade 3B with very broad tolerance.
* ISO 5198 covers testing at precision levels of accuracy. This is the most stringent test with the tightest tolerances.
* DIN 1944 *Acceptance tests for centrifugal pumps* is an older standard, still used by some manufacturers.
* API 610 *Centrifugal pumps for general refinery service* is a more general design-based standard in common use in the oil and gas industry.
* ISO 1940/1 is commonly used to define dynamic balance levels for pump impellers.
* ISO 10816 is used to define bearing housing or pump casing vibration.
* DIN 1952 and VDI 2040 are withdrawn standards, but are still in common use to specify methods of flow rate (q) measurement.

A2.5 Inspection and test plans (ITPs)

Pump ITPs closely follow the chronological manufacturing, assembly and test activities for the pump. They normally include, as a minimum, the following items.

Pump casing

* Material test certificates (to EN 10 204)
* Material identification records
* NDE results
* Record of casting defects, magnetic testing (MT) and repairs
* Hydrostatic test (normally at a maximum of 2× working pressure)

Pump shaft and impeller

* Material test certificates and heat treatment verification
* NDE tests as specified
* Record of casting defects, MT and repairs
* Hydrostatic test (normally at a maximum of 2× working pressure)

Pump shaft and impeller

* Material test certificates and heat treatment verification
* NDE tests as specified
* Dynamic balance certificate (a common level is ISO 1940 grade G6.3)

Assembled pump

- Completed technical data sheet
- Guarantee acceptance test results and report
- Painting records and report
- Pre-shipping documentation review

Some manufacturers will exceed these minimum requirements, others will not.

A2.6 Pump acceptance test procedure

The pump acceptance test is carried out in a purpose-built test circuit in the manufacturer's works. In practice, the layout of the circuit may be difficult to see as some of it is often underneath the test bay floor plates. Luckily, most test circuits follow a similar pattern; Figure A2.3 shows what to look for. Note that there are effectively two different parts: the basic circuit for the q/H test and an auxiliary suction control loop, which is connected for the purposes of the NPSH test. The circuit has suitable instrumentation to obtain the performance data; most pump manufacturers have a fully computerised data-logging system to process the data and display the results.

Circuit checks

The first source inspection task when witnessing pump tests is to check the circuit. Points to look for are as follows.

- Normally a 'shop' motor (i.e. not the contract motor) is used to drive the pump. Make sure it has the correct, or higher, power rating.
- Check the pipe arrangements either side of the flow rate measuring device: there should be a sufficient straight run in order not to introduce inaccuracies.
- Check the suction and discharge arrangements on either side of the pump. The pressure gauge or manometer connections should be at least two pipe diameters from the pump, or readings will be inaccurate.
- Watch for flow straighteners fitted before the pump. These are sometimes fitted to produce the required inlet flow characteristics, but they can produce pressure losses and distort the results.
- Ask the manufacturer to explain any variation of vertical levels throughout the circuit. These are particularly relevant to the NPSH test.

FIG A2.3
Centrifugal pump test circuit

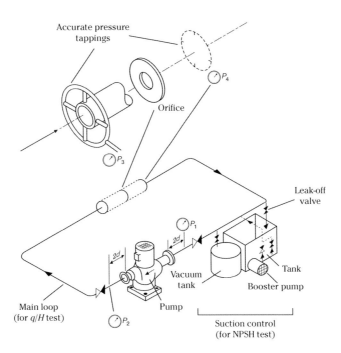

- Ensure that the volume of fluid in the circuit is sufficient to avoid temperature rise during the q/H test. If pump input power is high in relation to the volume, then additional cooling may be required.
- Calibration. Check calibration records for all measuring and recording equipment. Do not forget the transducers.
- Empirical factors. The pump manufacturer may have empirical correction factors built in to their calculation routines. Fluid density corrections and level corrections are two common examples. Check what they are.

Once circuit checks have been completed, the test then progresses in a fairly standard sequence of events.

Step 1. Obtain steady-state conditions

The pump is started and the circuit is allowed to attain steady-state conditions by running for at least 30 min. Use this period to make an initial check of the measuring equipment readings to ensure everything is working. Watch for any early indications of vibration or noise.

Step 2. The q/H test

The q/H characteristic is determined as follows. The first set of measurements is taken at the duty point (100%q). The valve is opened to give a flow rate greater than the duty flow (normally 120%q or 130%q) and further readings are taken. The valve is then closed in a series of steps, progressively decreasing the flow (moving from right to left on the q/H characteristic). With some pumps, the final reading can be taken with the valve closed, that is, the $q = 0$, 'shut-off condition'. This is not always the case, however; for high-power pumps, or those with a particularly high generated head, it is undesirable to operate with a closed discharge valve. During the test, it is useful to take particular care with the spacing of readings around the duty point, particularly for pumps where greater accuracy levels will be applied. Close spacing around the duty point will help the accuracy of the results by better defining the shape of the curve in the duty region.

Once the test points are obtained, you can now check them against the guarantee requirements. There are several discrete steps required here (refer to Figure A2.4).

- Draw in the test points on the q/H axes.
- Using the *measurement accuracy* levels given for the class of pump, draw in the q/H measured band as shown.
- Now add the rectangle, which describes the tolerances allowed by the acceptance guarantee on total head (H) and flow rate (q). Typical tolerances of $\pm 2\%H$ and $\pm 4\%q$ can be applied, if nothing is stated in the specification.
- If the q/H band intersects or touches the rectangle then the guarantee has been met (this is the situation in Figure A2.4). Note that the rectangle does not have to lie fully within the q/H band to be acceptable.

It is not uncommon to find different interpretations placed on the way in which pump standards specify acceptance tolerances. The standard clearly specifies *measurement accuracy* levels $\pm 2\%q$, $\pm 1.5\%H$), but later incorporates these into a rigorous method of verifying whether the

FIG A2.4
Checking *q/H* performance guarantees: Centrifugal pumps

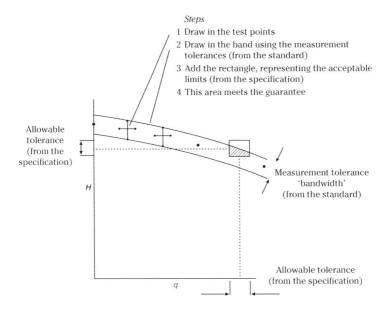

Steps
1. Draw in the test points
2. Draw in the band using the measurement tolerances (from the standard)
3. Add the rectangle, representing the acceptable limits (from the specification)
4. This area meets the guarantee

test curve meets the guarantee by using the formula for an ellipse (effectively allowing an elliptical tolerance 'envelope' around each measured point), specifying values of 2%H and 4%q to be used as the major axes lengths of the ellipse. Strictly, this is the correct way to do it – but the simplified method shown in Figure A2.4 is easier to use.

Step 3. The efficiency test

The efficiency guarantee is checked using the same set of test measurements as the q/H test. Pump efficiency is shown plotted against q, as in Figure A2.2. In most cases, the efficiency guarantee will be specified at the rated flow rate (q), the same one used for checking the head guarantee. The principle of checking compliance is similar: namely, draw in the characteristic bandwidth using the applicable measurement tolerances, followed by the rectangle representing any tolerances allowed by the acceptance guarantees.

Step 4. Noise and vibration measurements

Vibration levels for pumps are normally specified at the duty ($100\%q$) point. The most common method of assessment is to measure the vibration level at the bearing housings using the methodology of ISO 10816. This approximates vibration at multiple frequencies to a single velocity (RMS) reading. It is common for pumps to be specified to comply with a level of up to around 2.8 mm/s. Some manufacturers scan individual vibration frequencies, normally multiples of the rotational frequency, to gain a better picture of vibration performance. This does help with diagnosis, if excessive vibration is experienced during a test.

Pump noise is also measured at the duty point. It is commonly specified as an 'A-weighted sound pressure level' measured in dB(A) at the standard distance of 1 m from the pump surface. It can be difficult to obtain accurate noise readings during pump tests owing to the considerable background noise, which can come from turbulence in the rest of the fluid circuit. Any pump that has noise levels close to its acceptance noise level should be checked very carefully for excessive vibration levels, then particular attention should be paid to bearings and wear-rings during the subsequent stripdown.

Step 5. The NPSH test

There are two common ways of doing the NPSH test. The first is simply used for checking that the pump performance is not impaired by cavitation at the specified q/H duty with the 'available' NPSH of the test rig. This is a simple go/no-go test, applicable only for values of specified NPSH that can be built in to the test rig. It does not give an indication of any NPSH margin that exists, hence it is of limited accuracy. The more comprehensive and useful test technique is to explore NPSH performance more fully by varying the NPSH over a range and watching the effects. The most common method is the '3% head drop' method shown in Figure A2.5.

The test rig is used as for the q/H test, but with the suction pressure control circuit switched in (see Figure A2.3). The suction pressure is reduced in a series of steps and, for each step, the pump outlet valve is adjusted to keep the flow rate (q) at a constant value. The final reading is taken at the point where the pump head has decayed by at least 3%. This shows that a detrimental level of cavitation is occurring and defines the attained NPSH value, as shown in Figure A2.5. In order to be acceptable, this reading must be *less than or equal to* the maximum

FIG A2.5
Measuring NPSH: the 3% head drop method

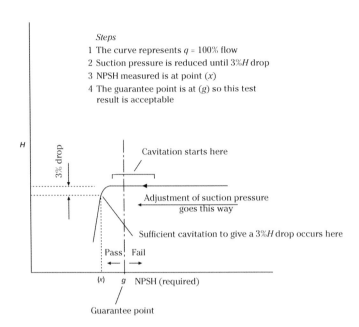

guarantee value specified. Strictly, unless specified otherwise, there is no acceptance tolerance on NPSH, although some standards give a *measurement* tolerance of ±3% or 0.15 m NPSH. Sometimes you will see this considered as being the 'acceptance' tolerance.

Corrections

There are a few commonly used correction factors that need to be used if the test speed of the pump does not match the rated speed, which often happens. The following factors will give sufficiently accurate results and can be applied to q, H, P and NPSH.

- Flow q (corrected) = q (measured) × (nsp/n)
- Head H (corrected) = H (measured) × $(nsp/n)^2$
- Power P (corrected) = P (measured) × $(nsp/n)^3$
- NPSH (corrected) = NPSH (measured) × $(nsp/n)^2$

where *n* is the speed during the test and *nsp* is the rated speed.

Step 6. The stripdown inspection

It is good source inspection practice to witness a stripdown inspection after the performance test. It is the inspector's best opportunity to check and report on some important design and manufacturing features of the pump. The pump standard API 610/ISO 13709 provides useful guidance on desirable mechanical design features. A typical pump stripdown checklist contains the following points.

- Observe the pump run-down; it should be smooth without any undue noise or unbalance.
- Check how the casing sections come apart; they should be a firm press fit, but should separate without needing excessive force.
- Check the casing joint faces for flatness; there should be no warping.
- Spin the shaft bearings by hand to check for any tightness or radial wear.
- Check the bearing surfaces; there should be no evidence of lubrication breakdown or overheating.
- Check the mechanical seals; any chipping or wear indicates incorrect assembly.
- Check the impeller fixing; it should be secured to the shaft with a cap nut so the spindle threads are not exposed to the pumped fluid. The impeller should have an acceptable fixing to the shaft (some specifications require a keyed drive, others do not).
- Check the wear-rings for excessive wear (get the limits from the manufacturer's drawings). The rings should be locked against rotation by a threaded dowel, not tack-welded.
- Surface finish is important. If in any doubt, use a comparator gauge to check for a finish of $0.4\,\mu m$ R_a or better on shaft and seal surfaces. Pump casings should have a finish better than $25\,\mu m$ R_a on outside surfaces and $12.5\,\mu m$ R_a on internal surfaces.
- Visually inspect the impeller water passages; smooth surfaces ($12.5\,\mu m$ R_a) indicate good finishing during manufacture. Look also for any evidence of the impeller having been trimmed, or 'underfilled' on the trailing edges, to make it meet its q/H requirements. These are acceptable, but only within limits.

All findings are recorded on the source inspection report.

A2.7 pump assembly checks

Witnessing of the pump assembly process can be a controversial issue – manufacturers do not always like the presence of an external inspector in their workshop when the assembly is in progress. This means that your access to witnessing the assembly activities fully is dependent on whether it is specified as a witness point in the ITP. From a purely technical viewpoint, the *need* for witnessed inspection of the assembly process depends on the type and size of the pump. Small pumps of straightforward design for use with lubricating oil, fresh water or condensate (namely the category commonly referred to by plant specifications as for 'general service' duties) are usually assembled in batches, from interchangeable parts, and can be considered low-risk items. In practice, it is rarely necessary or economical to witness the assembly of these items. The situation with larger, more specialised pumps is different and the following types of pump *should* have their assembly witnessed.

- Those using large-diameter castings (above about 600 mm).
- Vertical pumps with long extended drive shafts used for high-throughput seawater service (above about $0.5 \, \text{m}^3/\text{s}$), particularly if they have a rated speed of greater than 400 r/min.
- High-head pumps, including multi-stage centrifugal types used for boiler feed or borehole suction purposes.

These three categories of pumps have a much higher inherent technical risk than do general service types. They are more likely to perform poorly or to fail in early service if their mechanical assembly is not absolutely correct. The assembly of such pumps typically takes one or two days; this is sufficient time to check all the key dimensions and witness the important assembly steps. There are a couple of important points of principle to note about pump assembly checks. First, *focus* – the purpose of checking the pump assembly is to check those manufacturing activities that affect the successful future operation of the pump, rather than becoming involved in general discussion about the pump design or its non-critical pieces (interesting though they may be). Second, *reporting* – it is of little use witnessing the dimensional and assembly check only to record the results in a way that only you will understand. The results and data need to be treated carefully; the purpose is to record them so they can be used later as a 'baseline reference' of how things were done should the pump later fail in service or if any problems are discovered during the pump works test or stripdown examination.

We can now look at the common assembly checks, using a large vertical cooling water circulation pump for a power or process plant as an example.

Inspection of the cast components

Cast components should be visually inspected before assembly. Concentrate on the impeller and any of its attached rotating components rather than the stationary parts of the suction bell or discharge casings. Look for surface defects such as porosity or obvious cracking, particularly around areas of changes of section. It is not unknown for cracks to develop several days or weeks after heat treatment and NDE, as the stresses in the material relieve themselves fully. The commonly used standard for casting surface quality is MS-SP-55, which specifies acceptable numbers and sizes of surface defects and makes an attempt to address dimensional accuracy. Customers' specified requirements often provide a tighter standard to work to. SIs should not shy away from issuing a non-conformance report (NCR) if cast impellers exceed the MSS-SP-55 or purchase specification acceptance criteria; impellers are highly stressed components and surface defects can act as crack initiators, causing failure. For overall surface finish, look for a finish of 12.5 μm R_a as good practice for cast impellers.

Impeller blade profile

Accuracy of the impeller blade profile is important, particularly in large single-stage pumps. Differences in profile between blades have the effect of causing hydraulic unbalance; one blade will tend to produce significantly more (or less) pumping head than the others, causing a resultant radial force that 'precesses' at the same speed as the impeller rotation. The result is one-sided wear on the pump bearings, a condition that can easily cause the pump to fail. An inaccurate blade profile is mainly a result of the casting process used to manufacture the impeller. Large impellers are nearly always cast from sand moulds. This is a process that can introduce dimensional inaccuracies, even if the patterns themselves (which are usually computer numerically controlled (CNC) machined) are accurate. Ideally, impeller blade profiles should be checked with accurate steel or hard plastic templates, but these are not always available, except perhaps for large pumps. Another way is by using a multiple-axis coordinate measurement machine – an expensive and difficult exercise that is normally only done if specified as a contract requirement. In practice, during most source inspections, you will have no alternative but to do the best assessment you can *visually*. Concentrate on the following points.

- Check that all the blade leading edges look the same shape. There should be no irregular convex profiles caused by excessive hand-grinding.
- Measure the chord length of each blade as accurately as you can using a steel tape or ruler. Differences of more than about 1 mm are a cause for concern.
- Look at the blade tip machining; it should be of even profile all the way round the impeller, with no obvious steps or eccentricity.
- Check the profile of the blade undersurface; this may be almost flat or have visible undercamber, depending on the design of the pump. The important point is that all the blades should have the *same* profile with no visible differences. If you can see differences, issue an NCR.

Shafts

Shaft accuracy is particularly important for any pump with a shaft length, between the impeller fixing and motor drive coupling, of more than about 1000 mm. Vertical water circulation pumps may have several shaft sections joined together to give a total length of 10 to 15 times the impeller diameter. These shafts merit special attention during the assembly process. The main specified parameter is normally *concentricity*, as indicated by 'total indicated runout' (TIR); the measurement technique is basically the same as that shown for gas turbine rotors in Chapter 9. A typical pump shaft TIR limit is < 0.1 mm per 3 m length of shaft, but check the manufacturers' drawing for their imposed limits. If the shaft is outside its limits, it can be straightened by alternately heating one area of the shaft with a gas torch and then cooling with water from a hosepipe.

Shaft straightening can take several attempts, so some patience is required. Record the final TIR results at measurement points every 400 mm or so along the shaft, and at both ends. Note that shaft *straightness* is also important; concentricity by itself is not an accurate determinant of the truth of the shaft. As a rule of thumb, an acceptable shaft straightness is about 0.1 mm per 3 m length, the same value as for concentricity.

Clearances and fits

The correct clearances and fits are a basic but important part of a pump's assembly. The importance of obtaining accurate dimensions, particularly on bore diameters, increases with pump size. Typical guidelines are as follows.

- Bearing to shaft sleeve: running clearance fit
- Impeller to shaft: transition fit better than H7/k6. Check also that the bore diameter is truly parallel to these tolerances
- Casing wear-ring to casing: locational interference fit
- Impeller to casing wear-ring: running clearance fit.

A2.8 common non-conformances and corrective actions for centrifugal pumps

The following table lists some common non-conformances and corrective actions for centrifugal pumps.

Non-conformance	Corrective action
The q/H characteristic is above and to the right of the guarantee point (i.e. too high)	For radial and mixed-flow designs, this is rectified by trimming the impeller(s). The q/H curve is moved down and to the left. Watch for resultant changes in dynamic balance. Repeat the test
The q/H characteristic is 'too low' – the pump does not fulfil its guarantee requirement for q or H	Often, up to 5% head increase can be achieved by fitting a larger-diameter impeller. If this does not rectify the situation, there is a hydraulic design fault, probably requiring a revised impeller design. Interim solutions can sometimes be achieved by - installing flow-control or pre-rotation devices - installing upstream throttles
NPSH is well above the acceptance guarantee requirements	This is most likely a design problem, the only real solution being to redesign the pump and then repeat the test

Non-conformance	Corrective action
NPSH result is marginal	This can sometimes be a problem of stability. The right thing to do is to try the test again and see if it gives an exactly reproducible result, paying particular attention to the measurement of the 3% head decay (watch and listen for evidence of cavitation). It is sometimes possible to accept marginal NPSH performance under concession; to do this properly you need to check the system NPSH available to see whether a satisfactory pressure margin (about 1 m) still exists
Excessive vibration over the speed range	The pump must be disassembled. First check the impeller dynamic balance. Next check all the pump components for 'marring' and burrs – these are a prime cause of inaccurate assembly. During re-assembly, check concentricity by measuring total indicated runout (TIR) with a dial gauge. Check for compliance with the drawings, then repeat the test. Check the manufacturer's critical speed calculations. The first critical speed should be a minimum 15–20% *above the rated speed* – then do all the checks shown above
Excessive vibration at rated speed	It is important to describe carefully the vibration that is witnessed. High vibration levels at discrete, rotational frequency are a cause for concern. A random vibration signature is more likely to be due to the effects of fluid turbulence

Non-conformance	Corrective action
Noise levels above the acceptance guarantee levels	Pump noise is difficult to measure because it is masked by fluid flow noise from the test rig. Take this into account. If high noise levels are accompanied by vibration, a stripdown and re-test is necessary

Index

acceptance criteria of API 1104, 114–118
active excavation area, 151
anomalies, 128–130
   stable, 128
   time-dependent, 128
API 1169 Body of Knowledge (BoK), 19–37
   illustration, 25
API 1169 exam, 5, 6
API 1169 programme scope, 4
API 5L line pipe, 77–79
API 5L product specification levels (PSLs), 77
API 5L, grade X52, 77
API certificate examinations, 3
API exam question types, 6
   direct quote questions, 7
   elimination questions, 11
   'general knowledge' questions, 11
   paraphrase questions, 9
API ICP examinations, 5
   what to expect, 6–15
API Q1, 261–262
API RP 1169 inspection requirements, 38–51
API RP 1169 Job Safety Assessment (JSA), 40
ASME B31.4 and B31.8, 84–94
ASME B31.4 inspection and testing, 90–92
ASME B31.4 scope, 91
ASME B31.8, training verification, 92
ASME construction codes, what are they? 84–86
automatic orbital welding, 117

blasting and explosives, 199–202
blasting inspector, 45
brake testing, 206

bridges, 248

call before you dig, 149
cathodic protection (CP) corrosion control, 48
causeways, 248
CEPA/INGAA guide, 34, 52
CEPA/INGAA Practical Guide for Pipeline Construction Inspectors, 34–37
chief inspector, 45
code defect acceptance criteria, 114
Code Effectivity List, 1, 19
cold bending, 121
Common Ground Alliance (CGA), best practice document 13.0, 145–149
   pipeline locating and marking, 150
Common Ground Alliance (CGA), one-call system, 148, 149
   call before you dig, 149
Common Ground Alliance (CGA), stakeholder group icons, 148
content of API RP 1169, 26–34
corrosion inspector, 45

dams, 248
dikes, 248
dimensional tolerances, 82
ditching and excavation, 46
double-ending, 117
downhill welding, 279

encroachment area, 151
environmental contamination, 43
erosion control measures, 242
ERW pipe, 80
excavation tolerance zone, 151
explosives, road transport, 203

350 Index

fall protection, 209
federal pipeline regulations 49 CFR 192 and 49 CFR 195, 138–144
fire prevention and protection, 218
flammable liquids fire risks, 29 CFR 1926 subpart F, 209–212

gas pipeline location classes (ASME B31.8), 88
geofoam blocks, 285
grizzly bears, 279–281

hardness testing of repair welds, 107
hazardous materials table, OSHA 49 CFR 172, 215–216, 217
hazardous substances, 215–217
health and safety exam questions, 164–166
horizontal drilling inspector, 45

INGAA CS-S-9 pressure test safety, 125
INGAA pipeline crossing guidelines, 149
   active excavation area, 151
   encroachment area, 151
   excavation tolerance zone, 151
inspector health and safety responsibilities, 161–166
ISO 9000, 261–262

leak test, 129, 132
line cleaning and hydrostatic testing, 50–51
line marking, 144
lowering-in, 50

mallard ducks, 283–284
materials handling, 29 CFR 1926, 212–214
migratory species, 251
migratory bird permits, 50 CFR 21, 251
minimum clearance, 143
minimum coverage, 143

national pollution contingency plan, 40 CFR 300, 244–247
NDE acceptance criteria, 103
nominal pipe size, 79

operation of cranes and derricks, 202–208
OSHA 29 CFR 1910, 186
OSHA 29 CFR 1926, 189–191
OSHA health and safety regulations, 185–222

PHMSA, 275
pigging, 282
pigs, 281–283
pipe bending, 49
pipe handling and stringing, 46
pipeline and water body crossings, 47
pipeline crossing terminology, 152
pipeline dents, acceptance limits, 122
pipeline earthquake protection, 286
pipeline environmental impact assessment (EIA), 287–291
pipeline failure mode, 285
pipeline inspector, 30
   as quality manager, 260
   environmental responsibilities, 42, 235–237
   quality responsibilities, 259–264
   safety responsibilities, 39
pipeline locating and marking, 150
pipeline mapping – the NPMS, 275–276
pipeline pressure testing, 123–134
   INGAA CS-S-9 pressure test safety, 125
   leak test, 129, 132
   pressure ratio, 130
   safety requirements, 133
   spike test, 129, 131
   spike test, pressure reversal, 131
   strength test, 129, 130
pipeline surface distortions, 118–122
   buckles, 119, 120
   dents and bulges, 119
   gouges and grooves, 119
   knobs, 119, 120
   misalignment, 119, 120
   notches, 119
   out-of-roundness, 119, 120
   peaking, 119, 120
   rucks, 119, 120
   wrinkles, 119, 120
pipeline welding, 102

pressure reversal, 131
project planning participation, 32
protection of birds and endangered species, 249–251
protection of navigable waterways, 247–249
public exhibitions and demonstrations, 218

qualification of welders, 101
quality (QA) monitoring, 30
quality management, 259–261

repairs of pipeline spools, 142
rigging equipment requirements, 213
right of way (ROW) requirements, 46
road transport of explosives, 203
role of the pipeline inspector, 16–18, 28

safety representative, 32
scaffolding and fall protection, 208–209, 210
scaffolding certification, 208
seamless pipe, 81–82
shoring systems, 199
slider bars, 286
soil and land erosion, 241
soil classification types, 191–195
soil terminology, 194
specialist pipeline terminology, 270–275
spike test, 129, 131
spike test, pressure reversal, 131
stable rock, 193
strength test, 129, 130
surface condition, 82
sway-braces, 286

tag-out of equipment, 206
terminology, specialist pipeline, 270–275
The Endangered Species Act 1773, 249–250
traceability, 82
trench excavation, 191–199
trench safety hazards, 192
trench shields, 199
trench slope-breakers, 243

trench sloping and benching, 196–197

unstable rock, 193
uphill weaving technique, 279
upland erosion control, 243
  vegetation and maintenance plan, 241–244
upland regions, 241

ventilation, 218
verification of qualifications, 44

waterbody, 237
  major waterbody crossings, 238–239
welding and cutting safety, ANSI Z49.1, 218–222
  fire prevention and protection, 218
  public exhibitions and demonstrations, 218
  ventilation, 218
welding area fire protection, ANSI Z49.1, 220
welding code, API 1104, 95–108
  role of welding codes, 95
welding inspector, 45
welding PPE, 222
welding
  controlling documentation, 99
  destructive test specimen for PQRs, 100
  downhill welding, 279
  face bend test, 100
  hardness testing of repair welds, 107
  NDE acceptance criteria, 103
  of pipelines, 102
  qualification of welders, 101
  repair of weld defects, 105–107
  root bend test, 100
  side bend test, 100
  uphill weaving technique, 279
wetland and waterbody construction and mitigation procedures, 237–241
wetland crossings, 240
wrinkle bend, 141

50 CFR 21 Migratory Bird Permits, 251